在地球上，还没有哪种动物像它一样受到全人类的关注和宠爱。
谨以此书奉献给世界上所有爱着大熊猫以及将会爱上它的人们。

No other creatures on the earth have received as much care and love as it has from human beings.
This album is dedicated to those who love or come to love the giant panda.

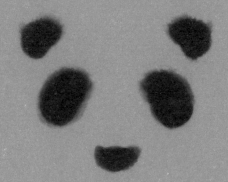

THE GIANT PANDA OF CHINA

周孟棋 摄影

Photographed by Zhou Mengqi

四川美术出版社

Sichuan Fine Arts Publishing House

大熊猫不仅是中国人民的宝贵财富
也是全人类珍视的自然历史的宝贵遗产

——世界自然基金会（WWF）宣言

The giant panda is not only the precious wealth of the Chinese people,
but also the precious heritage of the natural history cherished by all mankind.

— World Wildlife Fund (WWF) Declaration

目录
CONTENTS

大熊猫（拉丁文学名：Ailuropoda melanoleuca）属熊科，数量极其稀少，是世界上最珍贵的动物之一。目前，大熊猫仅生活在中国的特定地域，栖息地主要在四川、陕西、甘肃等周边山区。大熊猫被誉为"中国国宝"，它们穿越800万年，适应地质变迁和气候变化繁衍至今，成为动物界极为罕见的"活化石"。

2021年是科学发现大熊猫152周年。1869年，法国博物学家阿尔芒·戴维在中国四川进行科学考察时，第一次发现了大熊猫，并将其标本送回法国国家自然历史博物馆，首次将这一神秘物种介绍给了西方世界。正是在这百余年间，黑白相间、憨态可掬的大熊猫从中国的深山峡谷走向了全世界，进入人们的视野。

1985年，世界自然基金会（WWF）宣布大熊猫为世界十大濒危物种之一。经过中国政府的大力保护，2016年，世界自然保护联盟（IUCN）宣布将大熊猫受威胁等级从"濒危"降为"易危"。第四次大熊猫调查结果显示：中国大熊猫野生种群数量稳定增长。野生大熊猫种群数量达到1864只。其中，四川省有1387只，约占全国野生大熊猫总数的75%。截至2020年11月，大熊猫最新统计数据显示，全球圈养大熊猫种群数量达600只。其中，中国大熊猫保护研究中心圈养大熊猫种群数量突破300只。

数据让人振奋，这既是一份天赐的美好，也是对全人类为大熊猫繁衍生息所付出努力的回馈。作为最具辨识度的"中国符号"，大熊猫身上既蕴涵着中华民族优秀的文化精神——坚韧内敛、包容应变、顺势而为、友善和平，也契合了当前改善地球环境、维护生态平衡这一全人类共同主题。这一神奇的生物，牵引并融合成独一无二的"大熊猫文化"热潮，跨越时空、国界，搭建起中国与世界各国人民的桥梁。

■ FOREWORD

The giant panda (Ailuropoda melanoleuca), actually a member of the bear family, is one of the most precious animals in the world as it is extant with scarcity. Unique to China, the giant panda is praised as the "National Treasure of China" and deservedly belongs to the first-class protected animals in China. Now, it mainly inhabits mountains in such provinces as Sichuan, Shaanxi, Gansu, etc.

With a life history of eight million years on earth, the species of giant panda has experienced a variety of geological and climate changes, developing itself into a rare "living fossil" in the animal kingdom.

The year of 2021 marks the 152nd anniversary of the scientific discovery of the giant panda. In 1869, the French missionary Armand David firstly discovered the giant panda when he made a scientific investigation in Sichuan Province, and took the specimen of the famous black and white bear back to France, providing the West with the first impression of the mysterious species. It is just during the over 150 years that the cute and lovely giant panda with its black-white "suit" has walked out of high mountains and deep valleys in China and appeared in the world, becoming a global super star.

In 1985, the World Wildlife Fund (WWF) recognized the giant panda as one of the world's top ten endangered species. Since the Chinese government managed to conduct a series of effective protections, the International Union for Conservation of Nature (IUCN) announced in 2016 that the giant panda of China was downgraded from 'Endangered' level to 'Vulnerable' level in danger rating. In 2015, China's fourth investigation of giant pandas showed that the number of feral giant pandas in China no longer decreased, but steadily increased, with 1,864 wild giant pandas in the end of 2013. Among them, 1,387 live in Sichuan Province, covering about 75 percent of the total number of wild giant pandas in the country. As of November 2020, the latest statistics of giant pandas show that the global population of captive giant pandas is 600, among which the population of the captive giant pandas kept in China Conservation and Research Center for the Giant Panda exceeds 300.

These data are absolutely exciting, which can be treated as a heaven-sent gift for all mankind who are striving for the living and reproduction of giant pandas. As one of the most recognizable 'Chinese Symbols', the giant panda bears such cultural spirits of Chinese nation as being tenacious and introverted, tolerant and adaptable, updated, and friendly and peaceful. In addition, it fits the current universal theme, "improving the earth's environment and maintaining the ecological balance". In fact, this amazing creature has developed into a unique "Giant Panda Culture" which enjoys popular support across space and time, and also builds a bridge across national boundaries and races.

At present, the giant pandas that 'live abroad' in view of sojourn, rent, etc. are mainly distributed in 19 countries and regions in Asia, Europe, Oceania and the Americas. These pandas have had deep feelings with local people and won hundreds of millions of fans around the world. The giant panda is actually a unique medium, spreading and expressing both China's cultural confidence and peace and friendship to the world.

目前，以旅居、租养等各种途径"客居"在世界各地的大熊猫，主要分布在亚洲、欧洲、大洋洲、美洲的19个国家。这些"熊猫旅居地"的黑白精灵，和当地民众产生了深厚的感情，拥有全球数以亿计的粉丝。大熊猫文化成为一种独特的现象，传播与表达了中国的文化自信与对世界的和平友好。

2021年6月，由四川美术出版社出版的《中国大熊猫》（中英双语版）大型精品画册正式公开发行。画册记录了大熊猫的神秘身世、物种特征、分布规律以及人类保护、拯救、研究大熊猫的漫长经历。此外，画册还用大量篇幅展现了大熊猫旅居世界各地，与世界人民建立的深厚情谊。全书收录近300幅珍贵摄影图片，图片兼顾视觉性、艺术性、知识性、科普性、新闻性，全景展现大熊猫历经800万年，从时光深处走来，与人类渐行渐近、日益亲密、和谐共处的历程。

本书作者周孟棋，是"大熊猫故乡"人。他满怀对故乡与自然的赤子之爱，长达30余年，持续关注着大熊猫及其栖息地，以无比的耐心和爱心，捕捉到许多我们难得一见的大熊猫原生态珍贵瞬间。

在他的镜头中，有大熊猫极尽完美的黑白写真，富有情趣的细微动作，甚至可与人类情感交流的神奇灵性。在他的镜头中，大熊猫不只是生物学意义上的"活化石"，更是自然界里的高贵生灵。作者以温情而深沉的视角，解读出大熊猫与自然、与人类完美的交融，透射出生命最温暖的光芒。

作者长年热爱并执着地拍摄大熊猫及其栖息地，历经艰难困苦、惊险曲折，拍摄了上万张艺术作品，积累了无数珍贵镜头，这些精品成果，都将甄选并呈现于本书，为读者还原一个真实的大熊猫世界：大熊猫是地球文明进程的一个镜像，透过它，我们能够窥探古老的地球奥秘，并更加重视当下的生态文明。

2018年8月，在北京"首届中国大熊猫国际文化周"上，周孟棋先生被四川省人民政府评为首批"大熊猫文化全球推广大使"。从2019年开始，作者又开启了大熊猫旅居地"大熊猫和它的故乡"摄影巡展计划，探访旅居在全球19个国家和地区的中国大熊猫。因新冠疫情对全球的影响，目前只完成了澳大利亚、日本、泰国、俄罗斯等站的巡展。

这是一部用史料与现实立体呈现中国大熊猫的科研与繁育、保护与放归、文化与交流的影像记录，代表着物种存留对地球的深远意义，极具出版价值和现实意义。

通过此次出版，希望广大读者透过大熊猫的表象看到大熊猫深邃的文化内涵，了解和认知这一古老物种的生命价值。保护大熊猫及其栖息地，实际上就是保护生物多样性，保护人类赖以生存的地球，从这个意义上理解也就是保护人类自己——人与自然是共生共荣的生命共同体。

In June 2021, the elaborate picture album *The Giant Panda of China* (Chinese-English Edition), published by Sichuan Fine Arts Publishing House, was officially released. The album records the mysterious background, species characteristics, distribution of giant pandas and the long history of human beings in protecting, saving and studying giant pandas, as well as the deep friendship the giant panda has built with the people around the world when it lives abroad. The book contains nearly 300 precious photographic pictures, each of which is a combination of visual effect, art, knowledge, popular science and news, and displays a panorama of the history of giant pandas with eight million years, coming out of the depths of time and gradually approached human beings, getting to be intimate and harmoniously co-exist with the latter.

Zhou Mengqi, photographer of this picture album, is a native of Chengdu, Sichuan Province, the "Home to the Giant Panda". With deep love and care for his hometown and the nature, Zhou has spent more than 30 years in taking many rare and valuable pictures of giant pandas in the raw.

Through his lens, readers can see a perfect co-existence of the black and white of giant pandas, their fine and interesting movements, and even their emotional exchanges with human beings. In these pictures, the giant panda is not only a "living fossil" in the biological sense, but also a noble being in nature, whose perfect blend with nature and human being is best highlighted and interpreted thanks to the photographer's unique vision and tender feeling.

Zhou has experienced much hardship in taking thousands of photos of the giant panda, many of the wonderful in which are included in this album. Reading the photos, one will be brought to the true world of the giant panda: the giant panda is just a representative image in the process of earth civilization, through which we can explore the ancient mysteries of the earth and attach more importance to present ecological civilization.

In August 2018, Mr. Zhou Mengqi was awarded among the first "Global Promotion Ambassadors of Giant Panda Culture" by Sichuan Provincial People's Government at "First China Giant Panda International Culture Week" held in Beijing. Since 2019, Zhou has launched a "Giant Panda and its Hometown" photo tour to call on those Chinese giant pandas who are now living in 19 other countries around the world. So far, he has only completed the tours in Australia, Japan, Thailand, Russia, etc. due to the global impact of COVID-19.

By adopting historical data and real materials, this album constructs an image record of the scientific research and breeding, protection and release, culture and communication of Chinese giant pandas. It represents the profound significance of species retention to the earth, and enjoys great publishing value and practical implications.

This publication aims to help readers see the profound cultural connotation of the giant panda underneath its image, and understand and recognize the life value of this ancient species. To protect giant pandas and their habitats is actually to protect biodiversity and the earth on which human beings live. In this sense, it is just to protect human beings themselves-Man and nature co-exist and jointhly flourish.

穿越800万年漫漫风尘，守望每一季轮回，"活化石"大熊猫身上，记录着地球时空变迁的秘密。

Passing through a tough history of eight million years, the "living fossil" giant panda has witnessed the secret evolution of the Earth.

据科学的观点，大熊猫是晚中新世云南热带湿地森林动物的
孑遗种，是地球发展最新阶段——第四纪地质与生物变迁过
程中最好的见证者。

Scientists believe that the giant panda is the survivor of the
animals who lived in the tropical wetland forests of Yunnan
Province during the late Miocene, right witnessing the
geological and ecological changes of the Quaternary Period.

5000多年以来，大熊猫一直被视为中华大地上的祥瑞之物，是表达天意的神使。

For more than 5,000 years, the giant panda has been seen as an auspicious animal by the Chinese people, expressing the will of Heaven.

越过地球历史上最严酷的第四纪冰川期，始熊猫远在人类出现之前，就已经将生命的旗帜，高扬在世界峰巅……

Ailuaractos Lufengensis, or the earliest giant pandas, survived the ruthless Quaternary glaciers and appeared on the earth long before human beings, raising their flag of life to the summit of the world.

I 远古的祥瑞神兽
An Ancient Auspicious Deity Animal
踏上大熊猫DNA的探秘之旅
Tracing the history of the giant panda

如果把地球进化史比作一天的话，有人类的历史不过短短几分钟，而大熊猫显然是早得多的"泰斗"级见证者，被誉为动物界的"活化石"。它们曾广泛分布于黄河流域以南的大半个中国，以及缅甸、越南、泰国北部。

作为历史上大熊猫最广泛的分布地，中国早在三四千年前，就注意到这种奇特的生物，在很多典籍的记载中，大熊猫被不同时代的中国人赋予了很多名字。大熊猫既是猛兽，象征着"战无不胜"，又曾是和平的象征，作为两军求和言好的旗徽；其皮毛曾经是珍贵的贡品和药材；唐朝武则天时期，活体大熊猫还曾经被当作国礼赠送给日本。

西方人认识熊猫，始于19世纪法国传教士戴维（Armand David）。之后，一批又一批西方冒险家来到中国，通过各种途径，猎捕大熊猫活体，获得毛皮和标本，至此，"熊猫热"被引向全球，延续百年，至今仍然有增无减。

由于自然环境的变化和人类生活半径的不断扩大，大熊猫的生活圈一退再退，种群迅速消减，濒临危机。四川拥有目前世界数量最多的大熊猫种群、最佳大熊猫原生环境，被誉为"大熊猫故乡"。

卧龙自然保护区，是人类社会对大熊猫表达爱意的善举。联合国教科文组织已把它纳入"人与生物圈"计划之中，世界自然基金会（WWF）也不失时机地与中国合作研究如何保护大熊猫——卧龙因此成为全球关注的焦点。

卧龙自然保护区地理条件独特，地貌类型复杂，景型多样，气候宜人，还有浓郁的藏、羌民族文化。区内建有相当规模的大熊猫、小熊猫、金丝猴等国家保护动物繁殖场，有世界著名的"五一棚"大熊猫野外观测站，已建成国内重要的以单一生物物种为主的大熊猫博物馆。

Compared with the evolution of the earth, the history of human beings is much shorter. The giant panda appeared long before human beings and is thus known as the "living fossil". It had lived in most parts of China in the south of the reaches of the Yellow River, as well as in Myanmar, Vietnam and northern Thailand.

As the largest distribution of giant pandas in history, China had noticed this strange creature as early as three or four thousand years ago. In many ancient Chinese books, the giant panda was given many different names by Chinese people of different times. The giant panda belongs to fierce animals, a symbol of 'invincibility', and also a symbol of peace as used by the two armies in battlefront to seek peace. Its fur used to be a precious tribute and medicinal material. And in the period of Empress Wu Zetian in the Tang Dynasty (AD 618-907), a living giant panda had been given to Japan as a state gift.

Thanks to the French missionary Armand David of the 19th century, westerners got to know the giant panda. Since then, one group of Western adventurers after another came to China, leaving no stone unturned to capture living giant pandas for their furs and specimens. As a result, a global panda craze started and has lasted over 100 years so far.

Due to the changes in the environment and the expansion of human activities in modern times, the habitats and population of the giant panda shrank promptly. They are actually in danger of extinction. At present, Sichuan enjoys the largest number of panda groups and the best habitats for them, thus known as the "Hometown to the Giant Panda".

The Wolong Nature Reserve in Sichuan has been included in the "Man and Biosphere Program" sponsored by the United Nations. The World Wildlife Fund (WWF) also loses no time in co-operating with China in the study of how to protect the giant panda in Wolong which has turned to be the focus of international attention.

Besides unique geographical conditions and complex landscapes, the reserve boasts pleasant climate and various natural sceneries, as well as the rich cultures of the Tibetan nationality and the Qiang ethnic minority. Within it, sizeable sites for the reproduction of the giant panda, the lesser panda and the snub-nosed monkey were built. The world-famous field panda observation station '51 Tent' and the Giant Panda Museum are also located there.

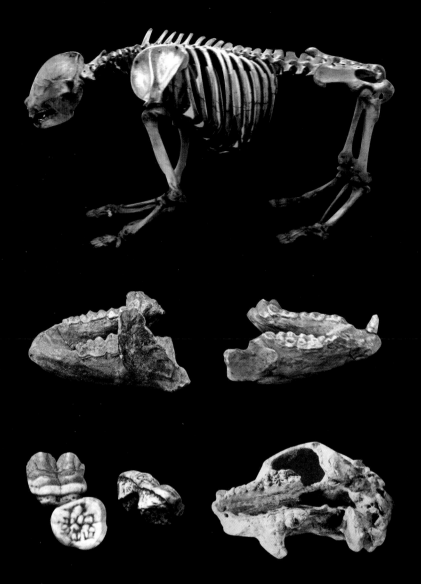

远古始祖，生物旗舰
Ancestor of the Giant Panda

在中国云南禄丰发现的大熊猫始祖化石，为破译大熊猫起源的千古之谜，提供了直接的证据。

在距今约800万年前的晚中新世时期，地球上就已经有了大熊猫的足迹。它们是大熊猫的先祖，它们的身体只有现在大熊猫的一半大，被科学家和考古学家命名为"始熊猫"。

The fossils of the earliest giant panda discovered in Lufeng, Yunnan Province in China offer direct proof for the origin of the giant panda.

During the late Miocene, about eight million years ago, the giant panda had appeared on the earth. As the ancestor of today's giant panda, it was half the size of the latter, called 'Ailuaractos Lufengensis' by scientists and archaeologists.

大熊猫历史分布图

Historical Distribution Map of the Giant Panda

生物学考古发现，早在远古时代的更新世，化石亚种大熊猫分布已相当广泛。

According to biological archaeological discoveries, as far as in the Pleistocene, the subspecies of the giant panda had been widely distributed.

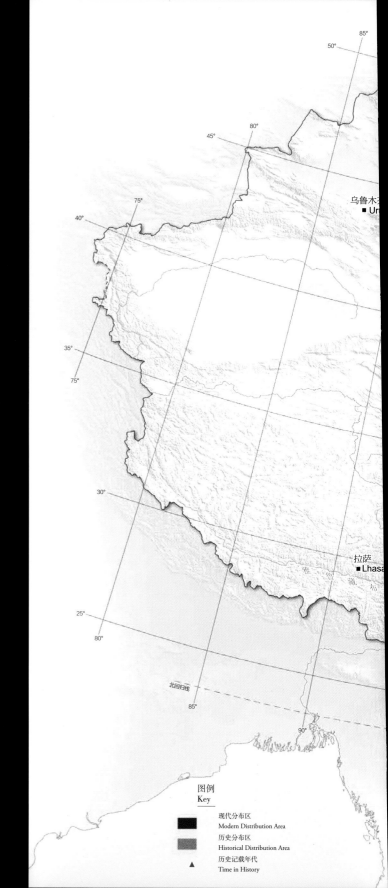

乌鲁木齐
■ Ur

拉萨
■ Lhasa

北回归线

图例
Key

现代分布区
Modern Distribution Area

历史分布区
Historical Distribution Area

▲ 历史记载年代
Time in History

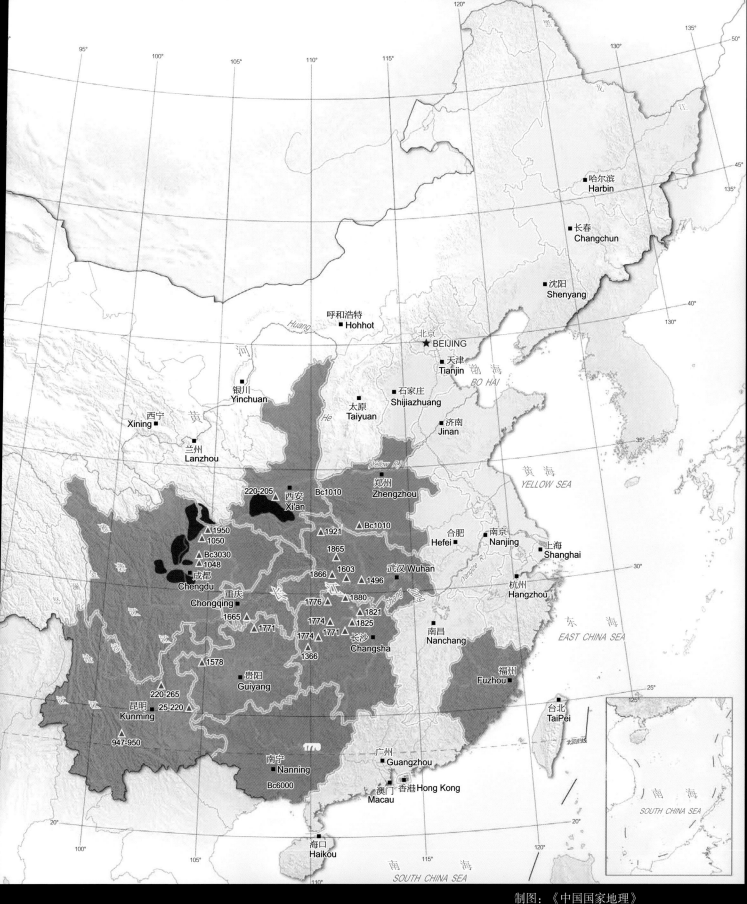

120°

95°   100°   105°   110°   115°   130°   135°

黑   50°

江

哈尔滨   45°
Harbin

长春   135°
Changchun

沈阳
Shenyang   130°

呼和浩特   北京
■ Hohhot   ★ BEIJING   40°

Huang   天津
河   ■ Tianjin   渤海
西宁   银川   BO HAI
Xining   Yinchuan   太原   石家庄
黄   Taiyuan ■ Shijiazhuang
兰州   He   济南   黄海
Lanzhou   ■ Jinan   35°   YELLOW SEA

Yellow R.   郑州
220-205   西安   Bc1010   Zhengzhou
△   Xi'an
1950 △   △ Bc1010   合肥   南京   上海
△ 1050   △ 1921   Hefei   ■ Nanjing   Shanghai
Bc3030   1865   30°
△ 1048   △ 1603   武汉 Wuhan
成都   1866 △ 1774 △ 1496   杭州
Chengdu   △   △   Yangtze R.   Hangzhou
重庆   1880
Chongqing ■   1776 △   △ 1821   东海
1665 △   1774 1771 △ 1825   南昌   EAST CHINA SEA
△ 1771   1774 △ 1771   长沙 Nanchang
△   △ 1366 Changsha
1578 △   福州
贵阳   Fuzhou
Guiyang   25°
220-265 △   台北
昆明   25-220 △   TaiPei
Kunming   北回归线
947-950 △

南宁   广州
Nanning   Guangzhou
Bc6000   澳门 香港 Hong Kong   南海
Macau   SOUTH CHINA SEA

20°   20°

海口   125°
Haikou

南海

南海
SOUTH CHINA SEA

100°   105°   110°   115°   120°

制图：《中国国家地理》
Graphics: CHINESE NATIONAL GEOGRAPHY

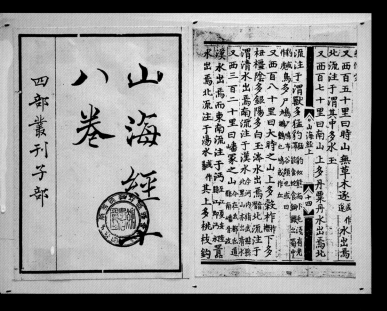

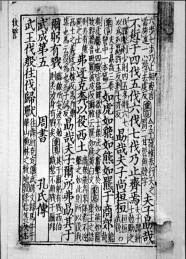

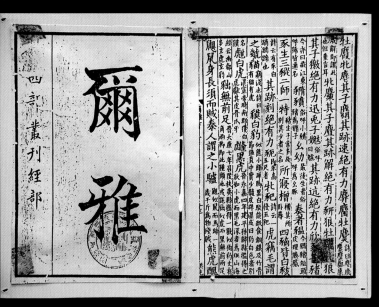

在中国历史上，大熊猫是一种传奇色彩浓厚的动物。早在3000年前的史料中，就有很多关于大熊猫的记载，仅古名就有十几个，如貘、貔、白熊、花熊、竹熊等。

In the history of China, the giant panda was a mystical animal. In the historical data of 3,000 years ago, there were many records of the giant panda which had a dozen ancient names such as tapir, white bear, striped bear, bamboo bear, and so on.

上：中国最古老的奇书《山海经》中的大熊猫让我们难以想象，那时的大熊猫牙齿锋利，能够嚼食铜铁，因此把它们称为"食铁兽"。

Upper: In the *Classic of Mountains and Rivers*, the oldest Chinese classic, the giant panda was thought to be enigmatic and called 'iron-eating beast', as people then thought its teeth were strong and sharp enough to chew copper and iron.

中、下：3000多年前的《尚书》和《诗经·尔雅》中描述说，貔貅（即大熊猫）像老虎一样威武，象征着战无不胜，它们的皮毛是向帝王进贡的珍品。

Middle and Lower: In the *Book of History* and the *Book of Songs* written more than 3,000 years ago, the giant panda was described to be as strong as the tiger, and treated as the symbol of the invincible. Its coat was used to be rare tribute to the imperial.

得一虎身文皆作飛鳥走獸之狀

戲嘴瓦屋山出貔貅常誦佛號子隴罌

雜州傳良選進士云其鄉蔡山多貔貅

嶺性食虎豹而馴于人常至僧舍索食

獨獸名似猱而大能食猿猿性羣獨獸

獨鳴一見五候鯖

宣和畫譜所載最古者吳曹弗興其玄

道碩按西京雜記漢元帝時有安陵陳

亮相震惊西方
Amazing Debut in the West

西方第一次发现大熊猫是1869年3月11日，法国传教士、法国国家自然历史博物馆标本采集者阿尔芒·戴维（Armand David），在四川省宝兴县的邓池沟第一次见到大熊猫皮后感到非常兴奋，他在日记中写道："在这条山谷中，一个李姓人家里，我见到一张展开的、那种著名的'黑白熊'皮，它非常特别，可能成为科学上一个有趣的新物种！"后来他把这张皮运回法国，并在法国国家自然历史博物馆展览，在欧洲引起了轰动。

On March 11, 1869, the French missionary Armand David first discovered the giant panda in Baoxing County, Sichuan and felt very excited. He wrote in his diary, "In the home of a person surnamed Li in a valley, I saw an unrolled coat of the famous black and white bear. It looks so special, expected to be an interesting new species in science." Later, he took the coat back to France and exhibited it in muséun national d'histoire naturelle in Paris, which caused a stir in Europe.

百年探秘热潮
Worldwide Exploration to the Giant Panda Lasting More Than a Century

西方探险家不断进入中国川藏地区寻找大熊猫，其中以美国西奥多·罗斯福总统的两个儿子克利姆特·罗斯福和小西奥多·罗斯福最为著名。他们于1929年组织了一次猎杀大熊猫的探险之旅，成为第一批猎杀大熊猫的西方人。在他们的回忆录《追踪大熊猫》中，他们这样描写："我们同时对渐行渐远的大熊猫背影开枪。两枪都命中……我们再次开枪，它应声而倒，但随后又爬起身，跑进浓密的竹林……"罗斯福兄弟及其他一些探险家猎杀的大熊猫后来成为在西方博物馆展出的第一批大熊猫标本，这在西方引发了一阵"熊猫热"，吸引了越来越多的探险家来华。

Interest in the giant panda prompted many Westerners to come to China. The most famous of them were Kermit Roosevelt and Theodore Roosevelt Jr., the two sons of former U.S. President Theodore Roosevelt. In 1929, they led an expedition team to hunt giant pandas in China, starting the Westerners' hunt for pandas. In their co-authored book *Trailing the Great Panda*, the brothers wrote "We simultaneously shot at the back of a giant panda that was walking away. Both shots hit it… we fired again and it fell down. But then it rose and ran into a dense bamboo forest..." The specimens of giant pandas killed by them and other explorers became the first exhibits of their kind in Western museums, which led to a 'panda craze' in the West, attracting more explorers to China.

比尔是一个富裕且具有冒险精神的探险家，他于1934年底前往中国捕捉大熊猫，但却意外地于1936年2月在上海去世。他新婚的太太露丝当时是纽约的一名时装设计师，从未有过任何野外探险的想法和经验。1936年4月，她决定到中国去取回丈夫的骨灰，并实现丈夫生前的愿望：将一只大熊猫带回美国……

Bill Harkness was a rich and adventurous explorer of the United States. He came to China at the end of 1934 to capture giant pandas, but unexpectedly died in Shanghai in February 1936. His newly married wife, Ruth Harkness, was then a fashion designer from New York, and had never thought of or experienced any field exploration. In April 1936, she decided to come to China to take back her husband's ashes and meanwhile realize his will of taking a giant panda back to the United States.

（以上三图选自成都大熊猫博物馆）
(The above three pictures come from Chengdu Panda Museum)

大熊猫栖息地
Habitat of the Giant Panda

雪山、林海、峡谷、溪流，风光绮丽的中国西部，青藏高原东部边缘——四川、陕西、甘肃的高山峡谷，大熊猫就隐居在这片如梦如幻的人间胜地。神奇的北纬30度线附近、海拔2600~3100米，这是北亚热带秦巴湿润区和青藏高原波密—川西湿润区的过渡地带，生物多样性区域。这些极为特殊的自然条件，为大熊猫创造了独一无二的生存乐园，成为庇护大熊猫免遭最后灭绝的"生命之舟"。

Giant pandas always live on a beautiful wonderland featuring snow mountains, forests, canyons and streams at the eastern edge of the Qinghai-Tibet Plateau, the juncture of Sichuan, Shaanxi and Gansu provinces in western China. The wonderland locates within a transitional area from the North Asian tropic humid zone to the western Sichuan humid zone. This area is near the mysterious 30 degrees north latitude, 2,600 to 3,100 meters above sea level, and thus enjoys great biodiversity. These special natural conditions create a unique living paradise for giant pandas, typically a Noah's ark to protect them from extinction.

山峦叠嶂，谷深林密，这种高山峡谷地势，是岷山山脉最为险峻秀美的地段。
With rolling mountains, dense forests and deep valleys, this area belongs to the most precipitous and majestic section of the Minshan Mountain Range.

世界自然遗产——四川大熊猫栖息地包括卧龙、四姑娘山、夹金山脉，总面积9245平方公里，涵盖成都、阿坝、雅安、甘孜4个市州的12个县市，是全球最大、最完整的大熊猫栖息地，全球30%以上的野生大熊猫栖息于此。四川拥有世界上数量最多的大熊猫种群、最佳的大熊猫原生态环境，被誉为"大熊猫故乡"。

Sichuan giant panda habitat includes the Wolong Nature Reserve, the Siguniang Mountain and the Jiajin Mountain. With an area of 9,245 square kilometers, the habitat covers 12 cities and counties within Chengdu, Aba Tibetan and Qiang Autonomous Prefecture, Ya'an and Ganzi Tibetan Autonomous Prefecture. It is the largest and most complete panda habitat in the world, with more than 30 percent of the world's wild pandas inhabiting. Sichuan Province sees the largest giant panda population and the best natural environment for them in the world, hailed as the "Hometown to the Giant Panda".

卧龙：天造地设，基因宝库
Wolong: Natural Gene Bank

卧龙国家级自然保护区地处四川盆地西缘、邛崃山脉东麓、成都平原向青藏高原过渡的高山峡谷地带，面积2000平方公里，享有"熊猫土国"和"亚洲大陆的生物广谱基因库"之美誉。

The Wolong National Nature Reserve is located between the western edge of the Sichuan Basin and the eastern foot of the Qionglai Mountain Range, just where the Chengdu Plain rises to the Qinghai-Tibet Plateau. Covering 2,000 square kilometers, the reserve is known as the 'Kingdom of the Giant Panda' and the 'Broad-Spectrum Gene Bank of Living Beings on the Asian Continent'.

九寨沟：人间仙境，隐居天堂
Jiuzhaigou Valley: Ideal Panda Habitat

大熊猫总是选择在最美丽的地方栖息。以"人间仙境"著称于世的九寨—黄龙地区，人类只能观赏和向往，大熊猫却能悠然安居。九寨沟位于面积达730平方公里的九寨天堂里，森林蜿蜒，海子众多，瀑布飞溅，钙华滩流，泉眼密布，整个景区宛若仙境。黄龙，让我们领略自然山川的极致美丽。

The giant panda loves to inhabit the most beautiful land. The Jiuzhaigou Valley-Huanglong area, known worldwide as an "Earthly Paradise," is only a tourist destination human beings yearn for, but a permanent domicile the giant panda can leisurely live in. Jiuzhaigou Valley is situated in the Jiuzhai Paradise which covers 730 square kilometers and boasts dense forests, a myriad of lakes, flying waterfalls and calcsinter springs, while Huanglong is where visitors can appreciate the most beautiful mountains and alpine lakes.

四姑娘山：蜀山皇后，生态乐园
Siguniang Mountain: Queen of Sichuan's Mountains and Ecological Eden

四姑娘山为邛崃山系主峰，由四座连绵不绝的山峰组成，与夹金山和卧龙同属邛崃山系，其最高峰幺妹峰海拔6250米，是四川第二高峰，有"蜀山皇后"之美誉。卧龙自然保护区就坐落在四姑娘山的东坡。

Together with the Jiajin Mountain and the Wolong Nature Reserve, the Siguniang Mountain belongs to the Qionglai Mountain Range, actually the highest peak in the range. It consists of four rolling peaks. Among them, the highest one is named Yaomei Summit, which is 6,250 meters above sea level, ranking the second highest among Sichuan's mountains, thus known as the "Queen of Sichuan's Mountains". The Wolong Nature Reserve is just located on the eastern slope of the Siguniang Mountain.

夹金山：宝山圣地，雄伟壮丽
Jiajin Mountain: Treasure Mountain with Majestic Scenery

在夹金山，人类第一次用现代科学的眼光发现了大熊猫。在天地与时空的流转中，夹金山犹如一尊庞大的守护神，守护着圣洁的远古生灵，让它们在静谧中从容生息，延绵至今。

It is just in the Jiajin Mountain that the giant panda was found by means of modern science for the first time. The mountain is like a great guardian angel protecting the cuddly bear, which can accordingly live quietly on the earth till today.

秦岭：地理坐标，自然屏障
Qinling: Geographical Coordinate and Natural Barrier

陕西佛坪国家级自然保护区，地处秦岭山系中心地带，雨量充沛，气候温和湿润，森林覆盖率达 95% 以上，是野生大熊猫分布密度最高的地方。

The Foping National Nature Reserve in Shaanxi Province is located in the heart of the Qinling Mountains, enjoying abundant rainfall, mild and moist climate, and a forest coverage rate of more than 95 percent. It is the most densely populated area of wild giant pandas in the world.

白水江：得天独厚，国宝云集
Baishuijiang: Blessed Land for Giant Pandas

甘肃白水江国家级自然保护区，横跨岷山和秦岭两大山系，山势由西北向东南倾斜，地形复杂，沟壑纵横，相对温差悬殊，气候和植被垂直分布明显，是我国大熊猫的重要栖息地。

The Baishuijiang National Nature Reserve in Gansu Province crosses the Minshan Mountains and the Qinling Mountains, sloping from northwest to southeast. The reserve is complicated in terrain, with crisscrossing valleys and gullies as well as a vertical distribution of vegetation, hence an important habitat for the giant panda of China.

成都：天府之国，熊猫家园
Chengdu: Heavenly Land and Home to the Giant Panda

成都，距野生大熊猫最近的中国城市。如境内的都江堰龙池国家森林公园，动植物起源古老，被中外专家誉为"野生植物基因库"；距成都最近的西岭雪山，终年积雪不化，富饶的原始森林覆盖率达90%，是动植物的天然乐园……这些地方也正是野生大熊猫的出没地。

Chengdu is the city nearest to wild giant pandas in China. Within it, the Longchi National Forest Park in Dujiangyan Irrigation System possesses a large number of ancient animal and plant lives, hailed as the 'gene bank of wild plants' by scientists at home and abroad; the Xiling Snow Mountain which is the nearest of its kind to Chengdu is covered with snow all year round, primitive forest coverage reaching 90 percent, regarded as a natural paradise of animals and plants. These regions are just haunts of wild giant pandas.

🐼 人与自然, 共荣相生
Peaceful Co-existence of Man and Nature

大熊猫栖息的原生态之地，也是藏、羌、彝等多民族聚居地。许多世纪以来，这里的人们一直对外界隐匿着这片世外桃源，他们与所有其他生灵一道，顺应着自然的指引，和谐地相伴生息。

The habitats in the raw of the giant panda are also where people of the Tibetan, Qiang and Yi nationalities live. For centuries, they have led a life in seclusion, enjoying peaceful co-existence with the nature.

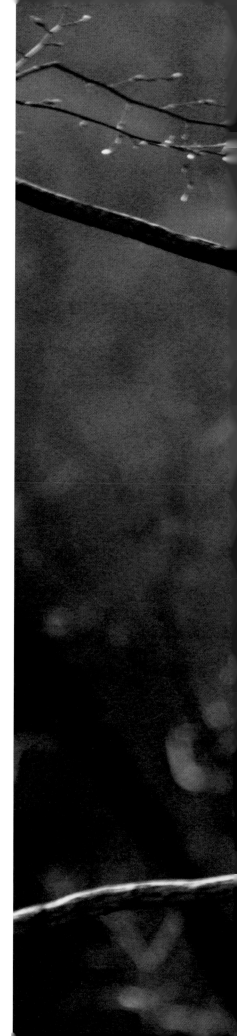

森林密友，和谐相伴
Fellows in Forest and Company in Harmony

这片亚热带山地的"绿色海洋"是大熊猫的栖息地，也是多数珍稀动植物的摇篮。金丝猴、扭角羚、小熊猫、毛冠鹿……它们曾与大熊猫共同经受住自然的多重考验，它们共受这片高山深谷的庇佑，和谐相伴，形成美妙和谐的生态家园。

This 'green sea' within the sub-tropical mountain lands is the habitat of giant pandas and also home for various rare species. Snub-nosed monkeys, budorcas taxicolor, lesser pandas, elaphodus cephalophus and other animals that had lived through various challenges of the nature together with giant pandas are enjoying the blessings of the mountains and valleys, harmoniously sharing this ecological Eden.

大熊猫的亲密伙伴——小熊猫，虽也以竹为生，却与大熊猫有着分享的默契。 ▲
The lesser panda is a close companion of the giant panda and also lives on bamboo.

大熊猫的亲密伙伴——扭角羚，是森林、灌丛和草甸之间的游民。美国著名动物学家夏勒（G.B. Schaller）说它是羚羊、牛、马、羊、骆驼和鹿"六不像"的独特动物。 ▲
The yakin, also a close companion of the giant panda, lives in forests, shrubs and meadows. According to G. B. Schaller, a famous American zoologist, the yakin is a unique animal, sharing the features of the antelope, the ox, the horse, the sheep, the camel and the deer in appearance.

大熊猫的亲密伙伴——金丝猴，中国国家一级"国宝"动物，毛色艳丽，动作敏捷，俗称"飞猴"，是飘荡在林海中的金色流云。 ▶
Also one of the close companions of the giant panda, the snub-nosed monkey is ranked among the 'State Treasure Animals' in China, thus enjoying first-class protection from the country. With shining golden fur, it always amazes its observers with elegantly agile movements, commonly known as 'flying money', just resembling a flowing cloud above the sea of forests.

Ⅱ 共荣共生的乐园
A Paradise of "Live and Let Live"

神奇的大熊猫世界
The mysterious world of the giant panda

只要见到它的模样，人人都会即刻被"降服"。黑白和谐，纯如处子；圆胖可爱，乖如婴孩；从容坚强，稳如长者。怎么能不爱上"三合一"的大熊猫？

从远古到现代的漫长穿越中，我们无法完全了解它经历了什么，但有一点是科学的共识：作为世界上最古老的生物之一，大熊猫能存活至今，它一定通过自己的方式，与"自然之母"灵性相通，达成非凡的默契。这些"非凡"，在它独一无二甚至神乎其神的各种生活习性中表露无遗：放弃肉食，转而食竹；迷恋清泉，河畔醉水；孤高自傲，独来独往；严择爱侣，呵护幼子；不畏严寒，冰雪勇士……如果说形态上的大熊猫无邪得让人忍俊不禁，那么性格里的大熊猫散发出的那种执着依循自然的光芒，则更令人敬畏——从形到质，大熊猫之美如此通透。

人们如痴如醉地渴望着了解那一个神奇的大熊猫世界。

As soon as one sees a giant panda, he or she will 'fall in love' with it. Black and white, it is as pure as a virgin; round and fat, it looks as cute as a baby; calm and firm, it appears as sedate as an elderly man. How can you resist the charm of a pure, cute and sedate creature?

We do not know what on earth the giant panda had gone through in its long evolution, but it is certain that the giant panda as one of the world's oldest creatures must have developed its unique way to adapt to the nature and so survived till today. In fact, the uniqueness can be clearly seen from its marvelous habits: giving up meat and eating only bamboo; loving springs and rivers; living like a lone hermit; being strict in selecting its partner, but very protective towards its babies; fearless of cold, ice or snow... By right of naiveté in appearance, the giant panda can make people simmer with laughter, while by right of trait of adapting to nature in personality, the creature can touch people with a sense of awe.

How we have a strong desire to approach the mysterious world of the giant panda!

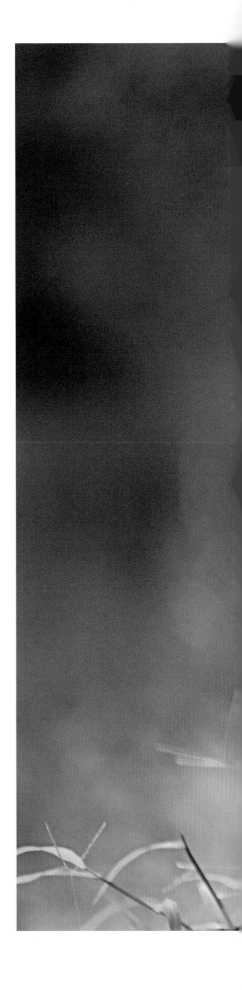

熊猫曲线
Round Giant Panda

圆脑袋、圆耳朵、圆眼睛、圆鼻头、圆身体。圆，构成了大熊猫完美的身体曲线。

Round head, round ears, round eyes, round nose and round body, the giant panda defines its cuteness with perfect round curves.

大熊猫是一种喜湿性动物，常栖息于高山深谷里的东南季风迎风面。这里气候温凉湿润，湿度常在80%以上。

Fond of humid setting, the giant panda prefers to live in the valleys facing southeastern monsoon, where the temperature is moderate and the air humidity is over 80%.

大自然造物的神奇，在大熊猫简洁明快的黑白之美、圆润乖胖的憨态上，体现得淋漓尽致。 ▶

The wonder of Nature in creating life is fully found in both the black-and-white beauty and the silly round appearance of the giant panda.

自然界没有绝对强大的统治者，适者生存的法则不应理解为攻伐或占有。从未"统治"过自然界的大熊猫，却能在时间的长河中自由徜徉。也许，大熊猫正是地老天荒的一封天书——敬畏自然，自然也庇护着它。

In its evolution, the giant panda has never tried to dominate the nature, but always enjoyed the peace and freedom in co-existence with the latter, vividly demonstrating the theory of 'survival of the fittest'. The history of the giant panda tells us an eternal secret of survival that only if in awe of the nature, one can be protected by it.

森林独居
Reclusion in Forests

多少世纪以来，大熊猫一直保持着独居的习性。它们独来独往，不群居，也不携偶，每只大熊猫都有自己相对孤立的巢域范围。

For many centuries, the giant panda has led a solitary life. It does not live in group, and refuses to live with its spouse. Each panda has its own independent sphere of activity.

雄性大熊猫每年活动范围约6~7平方公里，雌性大熊猫为4~5平方公里。这是它们独居的巢域。
A male giant panda has a territory of about six to seven square kilometers to live alone, while a female one about four to five square kilometers.

每只大熊猫都以自己独特的气味划定域界，不允许同性进入，偶有异性入侵，但极少见。
Each giant panda designates its territory with its own odor, forbidding other giant pandas of the same sex to enter the territory, even those of the opposite sex.

大熊猫活动的区域多在坳沟、山腹洼地、河谷阶地等，一般在20度以下的缓坡地带。这些地方土质肥厚，森林茂盛，箭竹生长良好，构成了一处气温相对稳定、隐蔽条件良好、食物资源和水资源都较丰富的优良食物基地。

The haunt of giant pandas is usually chosen in ravines, low-lying lands and river valleys, with gentle slopes of below 20 degrees. These regions enjoy fertile soil, dense forests and flourishing arrow bamboo, providing the giant panda with an ideal food and water-supplying base with stable temperature and safe shelter.

四川的自然保护区地处四川盆地向青藏高原急剧抬升的深山峡谷中，是中国保护高山生态系统及珍稀物种的最大绿色屏障。

The nature reserves in Sichuan are all located within high mountains and deep valleys which lie in the transitional zone from the Sichuan Basin to the Qinghai-Tibetan Plateau. These reserves construct the largest green screen in China which protects alpine ecological system and rare species.

河畔醉水
Intoxication of Water

大熊猫爱饮甘泉，常常"醉水"。只要遇到清澈甘甜的好水，就如人遇到好酒一样，必饮至肚子滚圆、不能走动，酷似一个贪杯的醉汉，久久躺卧溪边，最后才依依不舍地蹒跚离去。

The giant panda loves drinking spring water, often intoxicated with water. Once a giant panda runs into nice water, it will drink to its heart's content. After drinking too much water, it has to lie by the stream like a drunkard for a long time before staggering away.

大熊猫的家园总在清泉流水附近，以便就近畅饮泉水，即便到了严冬，它们仍要到溪流或不冻泉去饮水。如果家园附近没有水，它们会不惜长途跋涉到很远的山谷去寻水。

The giant panda always lives near clear spring so as to drink water easily. Even in severe winter, it will still go to drink water from unfrozen stream or spring. If no water is available near its habitat, the giant panda will have to travel a long distance to get water.

它们还是水中运动健将，涉水、渡水、游泳都很有一套。 ▶

Actually, they are good at such water exercises as swimming and wading across a river.

大熊猫有嗜水的习性，这与它的饮食结构有关。大熊猫所食竹类，其营养成分都是由可溶性营养素组成的细胞内含物，必须有充足的水分才能消化吸收，不能消化的大量木质素和纤维素也要靠大量水分才能排出体外。

The giant panda is addicted to water due to its diet. The nutritional composition of the bamboo it likes to eat is the cell inclusions consisting of soluble nutrients, which need to be digested and absorbed with plenty of water. A great amount of lignin and cellulose that cannot be digested by the giant panda also relies on a large amount of water to be eliminated from its body.

🐼 竹林隐士
Recluse in Bamboo Forest

到中新世，大熊猫已由食肉转变为以竹为主食，因这一奇特食性往往独栖于茂密的竹丛中，因此有了"竹林隐士"的雅号。它们最爱吃箭竹，一只成年大熊猫一天能吃竹叶10多公斤、嫩竹笋多达40公斤。也许是巧合——中国文化以竹象征正直与谦虚，大熊猫性情淡泊，恋守家园，它的品格与竹有关吗？

By the Pleistocene, the giant panda had gradually given up meat and mainly eaten bamboo. Due to this new eating habit, it has to live dispersedly in dense bamboo forest more often, hence the nickname 'Recluse in Bamboo Forest'. It enjoys eating arrow bamboo, and an adult giant panda is said to eat more than 10 kilograms of bamboo leaves and 40 kilograms of tender bamboo shoots per day. In Chinese culture, bamboo is the symbol of integrity and modesty, which coincidentally match well with the character and morals of the giant panda, since it really loves its home and displays indifference to fame.

研究发现，大熊猫由于朝着采食竹类的特化方向发展，它们的脑并不发达，长期隐居竹林，视觉也比较弱，但它们凭借灵敏的嗅觉和听觉弥补了不足。

Studies show that as the giant panda tends to specialize in collecting and eating bamboo, its brain is not developed at all. And the fact that it long lives in bamboo forest has led to its poor sense of sight, but fortunately its smell and hearing are good enough to offset the shortfall.

大熊猫的远祖是凶猛的肉食动物，现在虽然以食竹为主，但却依然保留着肉食动物的消化系统，这使它必须大量进食以摄取足够的营养。
With its ancestor fierce predator, the giant panda mainly eats bamboo, but still possesses the digestive system of former predator, which forces it to eat a lot in order to get enough nutrient substances.

因为以竹子为主食，它们的粪便甚至也是绿色的。

Its droppings even have a greenish color, the same as its staple food, bamboo.

大熊猫以食竹为主，日食量很大，因此每天必须花十多个小时奔波找寻粗嫩的竹笋、青翠的竹茎和茂盛竹子的枝叶以填饱肚皮。

The giant panda has a big appetite for its diet - bamboo, so each day it has to spend more than ten hours looking for rough but tender bamboo shoots, fresh culms and flourishing branches and leaves of bamboo to eat.

爬树勇士
Brave Tree Climber

大熊猫是爬树高手，在通常的情况下，它们爬树只是为了享受一次"日光浴"，或在树上眺望、嬉戏。有时，爬树同时又是求偶婚配和逃避敌害的一种方式。

The giant panda is good at climbing trees. Normally it climbs trees to enjoy a sun bath, or just overlook or frolic, but at times it aims to seek mating partners or evade stronger enemies.

高悬枝丫，大熊猫在树上更有安全感。◀
Clinging high to a tree branch, the giant panda feels much safer on trees.

大熊猫全身的关节十分灵活，可以像柔术演员一样完成各种高难度动作。▲
The giant panda grows very flexible joints, which helps it act like a contortionist.

横爬树干 ▲
Climbing across a tree branch

向上攀登 ▲
Climbing upward

三岁前的大熊猫特别喜爱爬树嬉戏，它们可以灵活地抱树打滚、倒挂，或者像玩滑梯一样溜下树，非常活泼好动。

Before three years old, the giant panda is particularly fond of climbing trees for fun. It is rather lively and active, and can hold a tree to neatly roll about or hang upside down, or just slid down a tree like on a sliding board.

🐼 穿越冰雪
Crossing Ice and Snow

当第四纪冰川期来临，气候骤冷，生存环境恶化，特别是食物缺乏，与大熊猫同时期的一些动物如巨猿、豺、中国犀、剑齿象相继灭绝了，唯有大熊猫退缩到部分高山深谷，活了下来，成为动物界的"遗老"、珍贵的"活化石"。

In the Quaternary Period, it abruptly became rather cold and the living environment turned worse. Due to the shortage of food, some of the giant panda contemporaries became extinct one after another, such as gigantic ape, jackal, Chinese rhinoceros and stegodon. Only the giant panda survived. It withdrew to high mountains and deep valleys, becoming the 'old die-hard' and rare 'living fossil' in animal kingdom.

对人类社会来说，它是珍稀的野生动物；对自然界来说，它是冰清玉洁的天之骄子。

For human society, the giant panda belongs to endangered animals, but for nature, it is God's favored one, quite pure and noble.

第四纪冰川期是生命的浩劫，大熊猫是造物者遴选出来的幸存者和勇士。
The Quaternary Period witnessed the destruction of life, and the giant panda was one of the lucky and brave survivors.

大熊猫身上的毛比较粗，毛里充塞的松泡髓质层是一种良好的保温材料，厚实的毛层表面还富含油脂，这一切都强化了其躯体的保温效应，甚至有祛风除湿的预防功效。

The giant panda grows thickset hair, in which the filled medulla substance is a fine heat preservation material. Besides, the surface of its thick coat is rich in oil. All these improve the heat preservation effect of its body, and even produce the preventive effect of dispelling wind and dehumidifying.

历经冰川严峻的考验，铸就了大熊猫不惧严寒的无畏性格，到了严冬也不冬眠。哪怕气温降到－14℃～－4℃，它们仍然穿行于覆着厚厚白雪的竹丛之中。

The experience surviving in the freezing Quaternary Period helps the giant panda never fear cold any more. It doesn't hibernate at all in severe winter. Even when it is 14 to 4 degrees below zero celsius, the giant panda can normally walk in bamboo forests laden with heavy snow.

大熊猫日活动高峰有凌晨和黄昏两次，特别是在冬季活动时间更长。一般凌晨三点即起，直至晚上七点活动高峰才结束。

The giant panda has two climaxes of activity each day, that is, before dawn and at dusk. In winter, it will act much longer, usually starting at 3 a.m. and coming to relax at 7 p.m.

与90万年前的化石相比，大熊猫仍然保持着古老的原始状态。
Compared with its ancestors living in 0.9 million years ago, the giant panda now maintains their primitive habitual nature all the same.

初春寻偶
Looking for a Partner in Early Spring

大熊猫性成熟后，每年发情一次，时间多在初春报春花、杜鹃花开的时节，一般为3月下旬至5月上旬。每年只有这个时候，它们才会从独居巢域中走出来，四处寻偶。

When a giant panda is sexually mature, it will be in rut once a year, mostly in the period when azalea and primrose are both in blossom in early spring. It is generally between late March and early May when a giant panda leaves its solitary territory to look for a spouse.

树上求偶

Conducting courtship in the trees

打斗是雄性大熊猫最公平的竞争方式。经过几个回合的较量，最后的胜利者成为新郎。新郎雄壮而不失风趣，新娘温柔而不失优雅，它们将在一起度过一段难忘的快乐时光。

Fighting each other is the fairest competition for male giant pandas. The winner will be the bridegroom. He is strong and witty enough and will, together with the tender and elegant bride, have a memorable and happy time.

正在交配的大熊猫　▲
The giant pandas in mating season

孕期中的大熊猫　▲
The giant panda during pregnancy

一般来说，如有机会，大熊猫在一个发情期内会与几个爱侣邂逅。它们每年选择的配偶并不固定，也不从一而终，而是较为开放，容易接受更多的追求者。　▶
Generally, a giant panda will encounter several partners in a rut if possible. It is not loyal to a fixed partner each year, but apt to mate with more pursuers.

形影不离
Love and care

丰富表情
Rich expressions

Ⅲ 熊猫工程：保护与回归
The Panda Project: Protection and Home Return

生命渴望回家
The giant panda desires to return its home in nature

　　20世纪七八十年代，由于环境变迁和人为因素，大熊猫栖息地遭到分割，形成一片一片彼此不相联系的"绿色孤岛"，导致野生大熊猫种群的缩小，这种远古生物正濒临危机边缘。因此，保护和抢救大熊猫成为全球的共识。

　　通过对大熊猫的人工繁育和适应性训练，将其放归大自然，从而达到延续该种群、使之与人类和谐共存的目的，这是人类保护大熊猫的有益尝试和大胆探索。中国大熊猫保护研究中心已拥有卧龙、都江堰、雅安三大基地。位于成都北郊斧头山的成都大熊猫繁育研究基地。经过30多年的艰苦努力，以6只大熊猫为基础，截至2020年，中国大熊猫保护研究中心圈养大熊猫种群数量突破300只，是世界上最大的人工繁育大熊猫种群基地之一。

　　世界各国的动物园和人们，对中国大熊猫的圈养、捐助不断，特别是汶川大地震后，大熊猫的灾后抢救、心理抚慰、基地重建，再次掀起了新一波熊猫关注热浪。

　　一件艺术品，即使被毁，还可能被构思出来，失传的乐曲，还可能被谱写出来，但当一种生物的最后个体停止呼吸，也许只有天地更新才会使它再现。每一种生物都有生存的权利，每一个生命都渴望着回家。回归自然，生命长存。

Due to environmental changes and human activities, the habitats of the giant panda were segmented, turning into isolated 'green islands', which resulted in the reduction of the giant panda's population. This ancient creature is actually at the edge of extinction. People worldwide have realized the importance and urgency of protecting and saving the giant panda.

Nowadays, the giant panda is artificially bred, then trained in adaptability, and finally released to the wild. This is a useful attempt and bold exploration to protect the giant panda, helpful to enlarge its population and make it harmoniously co-exist with human beings. China Conservation and Research Center for Giant Panda has so far constructed three bases in Wolong, Dujiangyan and Ya'an in Sichuan Province respectively. Located in the Futou Hill in the northern suburb of Chengdu, the Chengdu Research Base of Giant Panda Breeding originally possessed six giant pandas, but by 2020 it has ranked among the largest artificial breeding bases of giant panda in the world. the captive giant panda population in China Conservation and Research Center for Giant Panda exceeds 300.

An increasing number of zoos and people from different parts of the world are breeding the giant pandas from China by means of captivity or donating money. Particularly after the May 12 Earthquake in Wenchuan in 2008, a new round of attention was paid to the post-quake rescue, psychological comfort and base reconstruction of the giant panda in China.

A work of art can be reconstructed if destroyed, and a piece of music can be recomposed if lost. A plant or animal, however, cannot be reborn if dead. Every living thing has its right to survive, and each life is inclined to return to and mingle with nature.

成都大熊猫繁育研究基地是中国最早开展大熊猫长期国际合作研究的单位，先后与日本、美国、西班牙、法国、加拿大、德国、丹麦建立了"大熊猫长期国际合作繁殖计划"。目前，中国与近20个国家、20余家动物园开展了大熊猫保护合作研究项目。

Chengdu Research Base of Giant Panda Breeding is the first institution in China to carry out long-term international cooperative research on giant panda. It has successively established the 'long-term international cooperative breeding plan for giant pandas' with Japan, the United States, Spain, France, Canada, Germany and Denmark. So far, China has carried out cooperative research projects on giant panda protection with nearly 20 countries and over 20 zoos in the world.

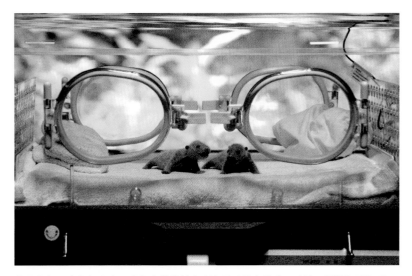

在北京奥运会期间出生于成都大熊猫繁育研究基地的大熊猫双胞胎"蜀祥""蜀云"
'Shuxiang' and 'Shuyun', the giant panda twins born in the Chengdu Research Base of Giant Panda Breeding during the Beijing Olympics in 2008

大熊猫新生幼仔呈粉色，娇小可爱。◀
A new-born giant panda has pink color, looking very petite and cute.

经过长期的艰苦努力，成都大熊猫繁育研究基地基本攻克了大熊猫繁育中发情难、配种受孕难和育幼成活难等几大难关，取得了举世瞩目的研究成果，创造了大熊猫繁育史的奇迹。

After long and arduous research, the Chengdu Research Base of Giant Panda Breeding has solved such difficult problems in the breeding of the giant panda as of estrus, mating and conceiving, and cub survival, making great research achievements which attract worldwide attention, and creating a miracle in giant panda breeding.

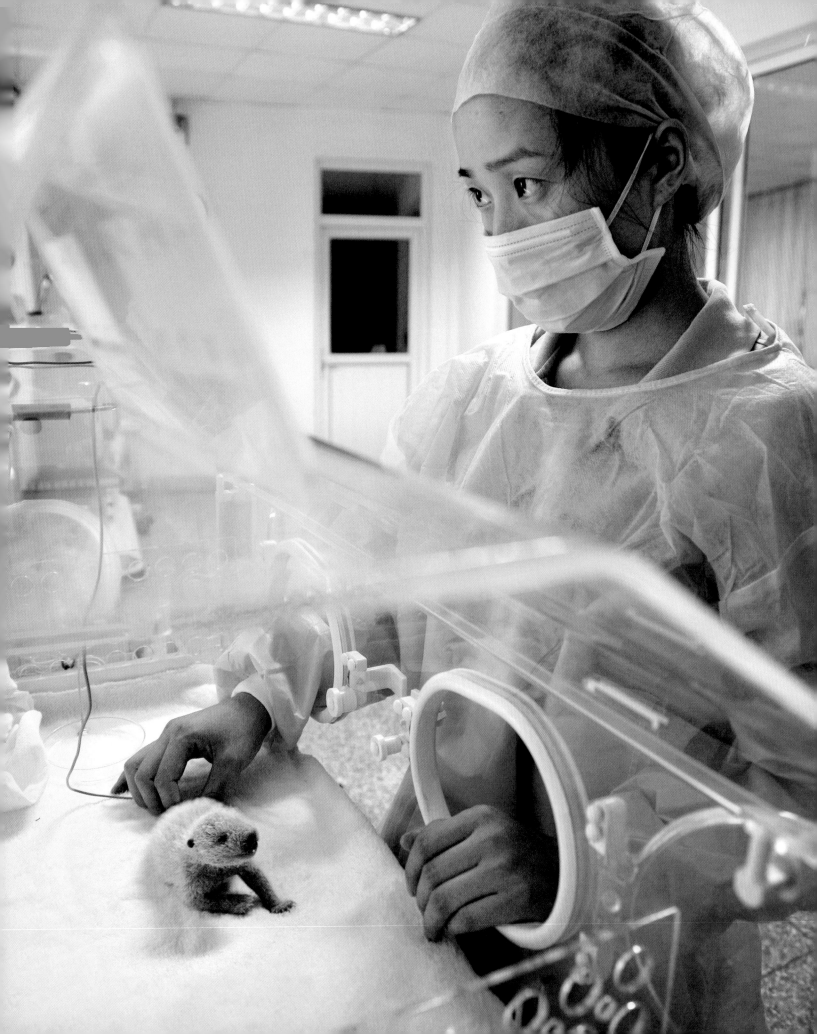

刚出生8小时，约120克
Just 8 hours after birth, about 120 grams

满月，约1100克
One month after birth, about 1,100 grams

一百天，约6000克
One hundred days after birth, about 6,000 grams

半岁，约16500克
Six months after birth, about 16,500 grams

40天，约1700克

40 days after birth, about 1,700 grams

两个月，约2900克

Two months after birth, about 2,900 grams

一岁

One year old

呵护

Good care

刚出生的大熊猫仅有成人手掌大小，体重仅为成年大熊猫的千分之一。它的体形从小到大的神奇变化，令人不可思议。20世纪80—90年代，圈养大熊猫幼崽成活率仅约30%；20世纪90年代以后，通过科研人员的长期努力与精心呵护，大熊猫幼崽的成活率提高到90%以上。

A new-born giant panda is only as long as an adult's palm, and weighs only one thousandth of an adult giant panda, which means the giant panda will see an amazing change in physique. In the 1980s and 1990s, only about 30% cubs in captivity could survive. Since the 1990s, the survival ratio has been increased to above 90% due to the long-time effort and good care of researchers.

黑白色彩之间，闪动生命最柔美的光芒。
Between the black and white, there shines the most beautiful light of life.

在幼仔出生后的一个多月里，大熊猫妈妈会一直把它抱在怀里。幼仔啼叫时，妈妈会不断舐它的身体，帮助它排便。待到幼仔断奶之时，大熊猫妈妈早已筋疲力尽。

For more than one month after its cub is born, the mother giant panda will cuddle the baby in her arms all the time. When the cub cries, the mother will keep licking its body and help it defecate. When the baby starts weaning, the mother will have always been exhausted.

大熊猫性情温顺，母性强烈。"熊猫小姐"如居闺阁深院的淑女，初次见人，常用前掌蒙面，或把头低下，一副娇羞模样；一旦它当上妈妈，爱子之情又十分强烈。

The giant panda is a creature of gentle temperament and strong maternal instinct. 'Miss giant panda' looks quite shy like a fair maiden who is still unmarried, and will cover her face with front paws or bashfully lower her head when first meeting a stranger. But when she becomes mum, she will have a strong love for her cub.

为使初生的幼仔在怀里感到舒适，大熊猫妈妈频繁变换坐姿。▶

To make the new-born cub comfortable in her arms, the mother giant panda has to change her sitting posture frequently.

这个世界如果没有大熊猫，如果没有
黑白二色，以及由黑白二色构成的
美，那将多么遗憾。

How sorry the world would be without
the giant panda, without the mixture
of black and white, and without the
beauty made of black and white.

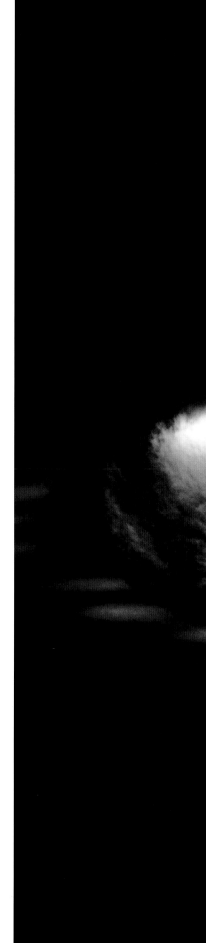

大熊猫一胎一般1~2崽。由于大熊猫母亲的精力有限，只能照顾一个宝宝，剩下的同胞兄弟姐妹只能人工喂养。

The giant panda usually gives birth to one or two cubs at a litter. But due to its limited energy, the mother giant panda can only take care of one cub by itself, so the other twin one has to be brought up by means of artificial feeding.

成都大熊猫繁育研究基地的科研人员在精心呵护大熊猫。
The staffs of the Chengdu Research Base of Giant Panda Breeding are taking good care of the endangered giant pandas.

为拯救大熊猫这一珍稀物种，中国大熊猫保护研究中心在保护栖息地的同时，逐步将圈养繁殖的大熊猫个体放归野外，以充实壮大大熊猫种群。

To save the giant panda from extinction, China Conservation and Research Center for the Giant Panda is gradually releasing the giant pandas bred in captivity back into the wild in order to enlarge the population of wild giant pandas.

很多时候，大熊猫简直与顽童无异，它们的许多动作，令人忍俊不禁。
Very often, the giant panda acts like an urchin, whose movements always make its visitors simmer with laughter.

大熊猫"琴心"
放归自然
The Giant Panda Qinxin
Reintroduction

保护的意义在于和谐共生，人工繁育只是人类抢救大熊猫的第一步，最终它们将会被放归山林，重返自然。自由，是灵魂最大的慰藉；生命，永远渴望回家……

The significance of protection lies in achieving harmonious coexistence. The artificial breeding is merely the first step for mankind to rescue the giant panda, which has to be released back into the wild in the end. Freedom is the greatest comfort of any soul, and for the giant panda, it always longs to return home and mingle with the nature.

IV 世界公民：交流与友谊
A Genuine Cosmopolite: Communication and Friendship

走向世界的大熊猫
The giant panda going global

　　7岁时，理查德·杰克逊第一次在华盛顿动物园里看到大熊猫。"那一天，好多人举着雨伞，排队观看从中国来的可爱动物"，这个情节成为理查德最为珍贵的童年记忆。

　　全世界的人都迷上了大熊猫，关爱大熊猫。

　　如果把"国宝"大熊猫出国的经历做个统计，相信频率不会亚于一位外交家。大熊猫的可爱与稀有，注定了它成为最受关注的"动物明星"。从"奥运会吉祥物"到"外交使者"，没有人料到，一个物种能以无与伦比的亲和形象，打破文化、政治、信仰的地域疆界，赢得世界人民特别是孩子们的喜爱，成为友谊与和平的化身。

　　大熊猫是最能代表中国"特产"的野生物种，也是最具辨识度的"中国符号"。目前，以旅居、租养等各种途径"客居"在世界各地的中国大熊猫，分布在19个国家和地区，这些"熊猫旅居地"的大熊猫，与当地民众建立了深厚的感情，拥有大批粉丝。通过大熊猫，人们也更加渴望了解中国文化。如今，大熊猫文化成为一种独特的现象，传播与表达着中国的文化自信以及对世界的友好姿态。

　　大熊猫，以它的独特方式，走向世界，传递美好情感。

At the age of seven, Richard Jackson saw the giant panda for the first time in the National Zoo in Washington D. C. He recalled, "That day, many people raising umbrellas queued up to watch the cute animal from China." It is the most precious moment Richard can ever remember about childhood. People worldwide are fascinated with the giant panda and care for it.

If counting the frequency of traveling abroad, the giant panda as 'national treasure' of China will be no less than a diplomatist. Being cute and rare, it is doomed to be the most focused 'animal star'. Acting as "the mascot of the Olympics" and "diplomatic envoy", the creature with its incomparable amiable image unexpectedly broke all the boundaries in culture, politics and religious belief, and won the love from people all over the world, especially children, becoming a symbol of friendship and peace.

The giant panda is the most representative wild species of China, and also one of the most recognizable 'Chinese symbols'. At present, the giant pandas living around the world in view of sojourn or rent are widely distributed in 19 countries and regions, where they have established deep friendship with local people and won a large number of fans. Due to the giant panda, people in foreign countries come to be more eager to understand Chinese culture and all about China. The giant panda culture has developed itself into a unique phenomenon which can spread and demonstrate the cultural confidence and kindness to the world of China as a great power.

In its distinctive way, the giant panda is going global toward a better future.

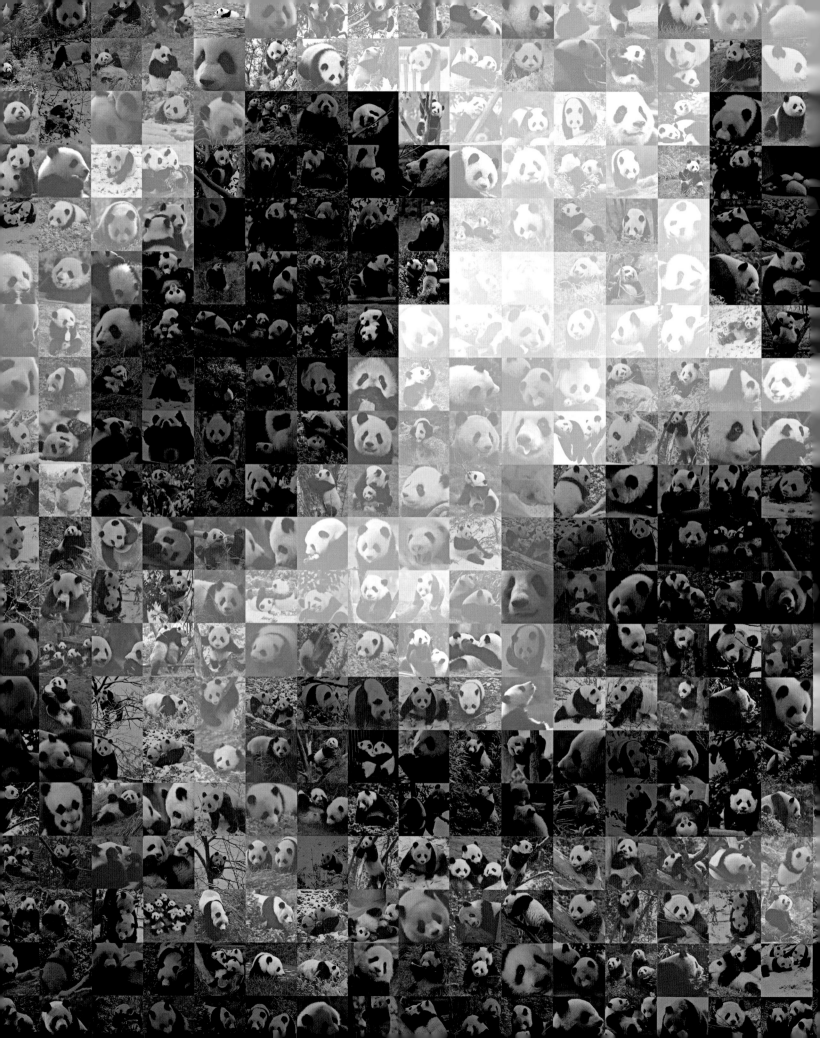

🐾 熊猫在世界
Giant Pandas Living around the World

大熊猫不仅是中国人民的宝贵财富，也是全人类珍视的自然历史的宝贵遗产。

——世界自然基金会（WWF）宣言

The giant panda is not only the precious wealth of the Chinese people, but also the precious heritage of the natural history cherished by all mankind.

— World Wildlife Fund (WWF) Declaration

中国 CHINA

日本 JAPAN

韩国 KOREA

马来西亚 MALAYSIA

新加坡 SINGAPORE

印度尼西亚 INDONESIA

亚洲
ASIA

泰国 THAILAND

大洋洲
OCEANIA

澳大利亚 AUSTRALIA

法国 FRANCE

英国 UK

 欧洲
EUROPE

美国 USA

比利时 BELGIUM

丹麦 DANMARK

德国 GERMANY

俄罗斯 RUSSIA

 美洲
AMERICA

加拿大 CANADA

西班牙 SPAIN

荷兰 HOLLAND

奥地利 AUSTRIA

芬兰 FINLAND

墨西哥 MEXICO

本版摄影、设计、插画：周孟棋、谢慧婕、殷顿
Photography, design, and illustration for this edition: Zhou Mengqi, Xie Huijie, Yin Di.

旅居世界，宠爱不衰
Staying Abroad as a Perfect Pet

大熊猫一次次远赴重洋，因外交途径、圈养与捐助而来到异国的动物园继续生活，这个"黑白配"的神奇生物无论在哪里亮相，都是人群与镜头的焦点。（图为美国亚特兰大熊猫馆）

In view of diplomacy, captivity or donation, the giant pandas of China embark on overseas journey over and over again to inhabit foreign zoos. The 'black and white' cuddly creature will always come under the spotlight as long as it appears. (The pictures show the Atlanta Giant Panda House in the United States)

在日本，人们要与大熊猫见次面，必须经历长达数小时的买票、入场等超长排队场面，然而等见到它，却只能停留2分钟，因为后面还有太多人！尽管这样，也挡不住千万"铁忠粉"的热情，他们乐此不疲，精神亢奋，如同盛大节日……

（左图为日本东京上野动物园，上图为日本和歌山白浜町动物园，下图为大熊猫香香）

In Japan, if to have a look at the giant panda, you have to spend several hours queuing up to buy a ticket and enter the zoo, but you can only stop for two minutes to watch it, as there are countless fans like you just behind! Even so, nothing can discourage millions upon millions of 'heartfelt fans' to be indefatigably involved in the festive visit of the giant panda. (The left picture shows the Ueno Zoo in Tokyo, Japan; the upper right picture shows the Shirahama Zoo in Wakayama, Japan; the lower right picture shows the giant panda 'Xiang Xiang')

英国爱丁堡动物园
Edinburgh Zoo in the UK

澳大利亚阿德莱德熊猫馆
Adelaide Giant Panda House in Australia

奥地利维也纳熊猫馆
Vienna Giant Panda House in Austria

比利时熊猫馆的饲养员接受媒体采访
The keepers of the giant panda house in Belgium interviewed by the media

泰国清迈动物园
Chiengmai Zoo in Thailand

2015年6月12日，法国、英国、德国、比利时、奥地利、西班牙6个国家选拔出的12名欧洲熊猫粉丝与大熊猫、南丝绸之路领域的专家和文化名人等组成的大型自驾团，驾驶"成都造"SUV，从欧洲出发，历时两个月，行经欧亚13国，行程两万多公里，向世界传播大熊猫和天府四川得天独厚的人文旅游魅力。

On June 12, 2015, a large self-driving group consisting of 12 selected European giant panda fans from France, Britain, Germany, Belgium, Austria and Spain, and the experts and celebrities in the fields of the giant panda and the South Silk Road drove the SUVs made in Chengdu and set out from Europe. Over a period of two months, they traveled more than 20,000 kilometers through 13 Eurasian countries to spread the culture of the giant panda and its hometown Sichuan with unique tourist charm to the world.

大熊猫粉丝在奥地利维也纳
Fans of the giant panda in Vienna, Austria

摄影 李浩
Photo by Li Hao

希腊国家旅游局官员在大熊猫推荐会上致辞
An official from the Greek National Tourism Administration addressing the giant panda recommendation

在德国举办的大熊猫图片展上留影的观众
Viewers at the giant panda photo exhibition held in Germany

欧洲熊猫粉丝不远万里来到中国探访熊猫故乡
European panda fans have traveled thousands of miles to China to visit the hometown to the giant panda.

各国小朋友热爱大熊猫
Children worldwide all love the giant panda.

波兰青少年与大熊猫玩偶合影
A group photo of Polish teenagers and a giant panda doll

我们喜爱大熊猫
We all love the giant panda.

我们喜爱大熊猫
We all love the giant panda.

与生俱来的非凡魅力，使大熊猫成为超人气明星。人们对熊猫的狂热不分年龄、种族与国界，分布在全球的大熊猫粉丝数以亿计，且与日俱增。"和平""友谊"与"爱"，从未在其他动物身上得到如此持久与深刻的体现。

The inborn extraordinary charm of the giant panda always makes it a super sta[r] about which people are all crazy regardless of age, race and national boundarie[s]. Around the world, there are hundreds of millions of giant panda fans, and th[e] number is daily on the increase. Such connotations as 'peace', 'friendship' an[d] 'love' attached to the giant panda have never been embodied so abidingly an[d] profoundly in all other animals.

在国际邮轮上欧洲熊猫粉丝向游客展示大熊猫摄影作品。
On an international cruise liner, some European giant panda fans are displayin[g] to the tourists the photographs of the giant panda.

世界小姐大赛佳丽齐聚"大熊猫故乡"——成都。
The beauties taking part in the Miss World Competition gathered and mad[e] their appearance in Chengdu, the 'Hometown to the Giant Panda'.

2019年4月，阿尔芒·戴维家乡的政府和家族后人不远万里来到雅安纪念大熊猫科学发现150周年。
In April 2019, the local government officials and families from the hometown o[f] Armand David traveled thousands of miles to Ya'an to commemorate the 150th anniversary of the scientific discovery of the giant panda.

 大熊猫纪事概览
Chronicle of the Giant Panda

800万年漫长的时光轴，刻录下大熊猫从远古走到今天的历史发展、自然变迁，亦记
载着人类科学发现大熊猫的曲折历程、保护实践，更续写着大熊猫与人类相依于地球的
深厚情谊、和谐美好。愿此记录，绵绵延续，为人类生命共同体谱写新的篇章。

Besides the historical evolution and natural vicissitude of the giant panda from the immemorial to the present, the long time of eight million years has recorded the tortuous process in discovering it and its protection in virtue of human science, and furthermore the far-reaching friendship and harmonious coexistence between the giant panda and human beings on the earth. We hope that this record will continue and contribute to the construction of a Commnity of Shared Future for Mankind.

- 800万年前的晚中新世，中国云南禄丰等地，生活着大熊猫的祖先——始熊猫（Ailuaractos Lufengensis）。

- 200多万年前的更新世早期到100万年前的更新世中晚期，大熊猫广泛分布于我国南半部，组成了大熊猫—剑齿象动物群。

- 2100多年前春秋战国时代的《山海经》中记载，大熊猫像熊，毛色黑白，产于邛崃山严道县（今四川荥经县），并说它食铜铁，故称"食铁兽"。

- 685年10月22日，唐朝女皇武则天将一对活体白熊（大熊猫）和70张皮作为国礼，送给日本天武天皇。这是迄今为止史书记载的第一次大熊猫出使。

- 1869年3月11日，在雅安宝兴邓池沟的法国天主教传教士阿尔芒·戴维（Armand David）第一次看到一张黑白相间的奇特动物皮，当时便估计是一种新物种。4月1日，戴维将猎人捕获的一只活体"竹熊"定名为"黑白熊"，归属于熊科。戴维是第一个向西方世界介绍中国大熊猫的外国人。

- 1871年，动物学家、法国国家自然历史博物馆馆长米勒·爱德华兹进一步鉴定，认为大熊猫属猫熊科，将它定名为"猫熊"。

- 1936年11月9日，35岁的纽约女服装设计师露丝·哈克尼斯在汶川县草坡乡捕获一只1月龄的大熊猫幼崽。这只取名"苏琳"的大熊猫被带到美国，在芝加哥市布鲁克菲尔德动物园展出时引起巨大轰动。在大熊猫的国际谱系册中，"苏琳"的编号是001。

- 1938年，在重庆北碚平民公园举办了一次动物标本展览。"猫熊"字样因书写顺序被误读成熊猫（Panda），从此人们也就习惯了这种叫法。

- 1938年，美国通过芝加哥《每日新闻》驻成都记者斯蒂尔（A.T.Steele），在四川获得一只熊猫"美兰"，送往芝加哥动物园展出。"美兰"一直活到1953年9月。

- 1953年1月，成都动物园从灌县（现都江堰市）获得一只约半岁的大熊猫幼崽，成为中国正式饲养展出大熊猫的动物园。

- 1957年，大熊猫"平平"是中华人民共和国成立后第一只被作为国家礼物送往苏联的大熊猫。

- 1959年4月1日，国宝大熊猫的形象出现在（特31）中央自然博物馆的邮票上。

世界自然基金会（WWF）会徽
Emblem of the World Wildlife Fund (WWF)

1980年6月，中国大熊猫保护研究中心在四川卧龙自然保护区建立。
In June 1980, China Conservation and Research Center for the Giant Panda was set up in the Wolong Nature Reserve, Sichuan.

1987年3月，建立了成都大熊猫繁殖研究基地。
In March 1987, the Chengdu Research Base of Giant Panda Breeding was set up.

位于秦岭中段南坡的陕西汉中大熊猫自然保护区（又称佛坪自然保护区），被称作是大熊猫在地球上最北缘的栖息地。
The Hanzhong Giant Panda Nature Reserve in Shaanxi Province, also known as the Foping Nature Reserve, is located in the middle and southern slope of the Qinling Mountains, referred to as the northernmost habitat of the giant panda on the earth.

中国大熊猫研究的标志性人物——西华师范大学生命科学学院胡锦矗教授在大熊猫实验室对大熊猫头骨化石进行研究。

As a forerunner in the giant panda research in China, professor Hu Jinchu from the School of Life Science of China West Normal University is studying the skull fossil of a giant panda in the giant panda laboratory.

中国大熊猫保护研究中心党委副书记、常务副主任张和民在野外考察。

Zhang Hemin, Deputy Secretary of the Party Committee and Deputy Director of China Conservation and Research Center for the Giant Panda, is on a field study.

中国大熊猫繁育技术委员会主任、大熊猫国家公园成都管理局书记张志和（右一）在实验室工作。

Zhang Zhihe (First right), Chairman of the China Technical Committee of Giant Panda Breeding and Secretary of the Party Committee of the Giant Panda National Park Administration in Chengdu, is working in the laboratory.

- In the late Miocene about eight million years ago, Ailuaractos Lufengensis, ancestor of the giant panda, lived in Lufeng and others, Yunnan Province, China.

- From over two million years to one million years ago, the giant panda was widely distributed over the southern half of China, forming part of the giant panda-stegodon fauna.

- According to the *Classic of Mountains and Rivers* more than 2,100 years ago, the giant panda looked like a bear whose coat was black and white, and lived in Yandao County (today's Yingjing County) in the Qionglai Mountain of Sichuan Province. The giant panda in the book is said to be an 'iron-eating beast' as it 'chewed copper and iron'.

- On October 22, 685 A.D., Empress Wu Zetian of the Tang Dynasty (618-907 A.D.) sent a pair of living giant pandas and 70 panda furs as state gift to Mikado Tianwu, the emperor of Japan then. It was the first recorded diplomatic mission for the giant panda of China.

- On March 11, 1869, French missionary Armand David encountered a peculiar black and white animal fur in Dengchigou, Baoxing County in Sichuan Province for the first time, and then considered it to be of a new species. On April 1, he named the living 'bamboo panda' captured by a hunter the 'black and white bear', and labeled it as Ursidae. David was the first foreigner who introduced the giant panda of China to the West.

- In 1871, Miller Edwards, zoologist and curator of the Natural History Museum in Paris, further identified the giant panda as a member of family Ailuridae and named it Ailuropoda melanoleuca.

- On November 9, 1936, Ruth Harkness, a 35-year-old clothing designer from New York, captured a 30-day-old giant panda cub in Caopo Township of Wenchuan County, Sichuan Province. The giant panda named 'Su Lin' was then brought to the United States and exhibited in Chicago Zoological Park, which created a great sensation in the country. Su Lin is 001 in serial number in the international genealogy of the giant panda.

- In 1938, an animal specimen exhibition was held in the Pingmin Park in Beibei of Chongqing, China, in which the giant panda's specimen was included. 'Maoxiong', the originally Chinese term for the creature, was mistakenly written as 'Xiongmao', the panda, which has remained ever since for panda in Chinese.

- In 1938, A.T. Steele, a U.S. resident reporter of Chicago *Daily News* in Chengdu, managed to obtain a giant panda named 'Mei Lan' in Sichuan Province. He then sent the panda to the Chicago Zoo, where it was exhibited. Mei Lan died in September 1953.

- In January 1953, the Chengdu Zoo got a giant panda cub about half a year old from Guanxian County (today's Dujiangyan), thus becoming the first zoo in China to raise and exhibit the giant panda.

- In 1957, the giant panda named 'Ping Ping' was sent to the former Soviet Union. It is the first time for the giant panda to be sent to a foreign country as state gift after the founding of the People's Republic of China.

- On April 1, 1959, the image of the giant panda appeared in the stamp 'Central Natural Museum' with the number T31.

- 1961年，大熊猫的形象被选为世界自然基金会（WWF）的会徽。

- 1963年9月9日，北京动物园首次开创人工饲养下成功繁育大熊猫的历史。

- 1963年3月18日，中国政府在四川汶川县建立卧龙自然保护区，面积2000平方公里。

- 1963年至今，在秦岭、岷山、邛崃山、大相岭、小相岭和凉山六大山系，先后建立了56个"大熊猫自然保护区"，对大熊猫密集的地区和栖息地实施有效的保护。

- 1964年，戴维斯（D.D.Davis）通过对美国动物园历年保存的大熊猫标本的研究，出版《大熊猫的形态学与进化机理的研究》专著。

- 1970年，中华人民共和国出借大熊猫给美国和日本的动物园，这被视为是新中国与西方国家初次的文化交流，是当时中国外交的重要环节。

- 1972年4月，中国赠送美国一对大熊猫"玲玲"和"兴兴"，"熊猫热"迅速席卷美国，当年成为美国的"熊猫年"。

- 1974年—1977年，四川、陕西、甘肃三省开始对大熊猫等珍稀动物进行第一次调查，统计大熊猫共2400多只。

- 1975年，岷山地区箭竹开花，大熊猫死亡138只以上；1983年岷山和邛崃山的冷箭竹大面积开花，灾后发现大熊猫尸体108具，抢救无效死亡33只，共计141只。

- 1978年9月8日，北京动物园大熊猫"娟娟"一胎产下两崽，这是世界上第一次为大熊猫人工授精获得成功的记录。

- 1978年，四川卧龙自然保护区建立了世界上第一个大熊猫野外生态观察站——"五一棚"观察站，中外科学家采用无线电跟踪等手段，对大熊猫个体生态、种群以及大熊猫主食竹类进行研究，取得了可喜的成果。

- 1978年，中国自然科学基金会与联合国珍稀濒危动物保护中心合作，成立了大熊猫联合保护科考小组。

- 1980年，世界自然基金会（WWF）受中国政府邀请参与中国保护野生大熊猫的行动中。

- 1980年6月，中国保护大熊猫研究中心在卧龙自然保护区建立。

- 1980年，卧龙自然保护区加入联合国教科文组织"人与生物圈"保护区网。

保护大熊猫世界名人签名活动得到国际组织和五大洲部分国家元首、世界知名人士的广泛响应和支持。
The initiative to seek signatures of celebrities in the world for the protection of the giant panda has drawn a positive response and support from many international organizations, heads of the states and worldwide celebrities of the five continents.

当大熊猫与人类温情拥抱，还有什么障碍不可逾越？
When we warmheartedly embrace the giant panda, no barrier will exist among us.

全球多家媒体争相报道在美国亚特兰大出生的大熊猫 "美兰" 周岁庆典的盛况。

Many media from different parts of the world covered the first birthday celebration of the giant panda 'Mei Lan' that was born in Atlanta, USA.

在日本，越来越多的成人和孩子爱上了这个胖乎乎的戴着 "墨镜" 的朋友。

In Japan, a growing number of adults and children come to fall in love with this chubby friend wearing 'sunglasses'.

中外游客争相观赏大熊猫。

Chinese and foreign tourists flock to watch giant pandas.

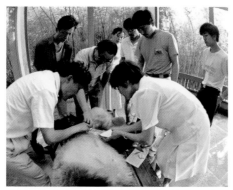

成都大熊猫繁育研究基地组织专家抢救野生病饿的大熊猫。

Experts from the Chengdu Research Base of Giant Panda Breeding in Sichuan Province are saving a wild giant panda that is sick and hungry.

- In 1961, the image of the giant panda was applied in the logo of the World Wildlife Fund (WWF).

- On September 9, 1963, the Beijing Zoo became the first in history to succeed in breeding the giant panda in captivity.

- On March 18, 1963, the Chinese government set up the Wolong Nature Reserve in Wenchuan County of Sichuan Province, which covers an area of 2,000 square kilometers.

- Since 1963, 56 nature reserves for the giant panda have been set up successively in the six mountains of Qinling, Minshan, Qionglai, Daxiangling, Xiaoxiangling and Liangshan, aiming to provide efficient protection to the regions or habitats where giant pandas live densely.

- In 1964, D.D. Davis had his book *The Morphology and Evolution Mechanism of the Giant Panda* published after studying the panda specimens kept in American zoos.

- In 1970, the Chinese government loaned giant pandas to the zoos of America and Japan, which was considered as the first cultural exchanges between the People's Republic of China and Western countries, regarded as an important part of Chinese diplomacy at the time.

- In April 1972, the Chinese government presented two giant pandas 'Ling Ling' and 'Xing Xing' to the United States as a gift, which immediately triggered a panda craze there, and this year was said to be the 'Year of the Giant Panda' of the country.

- From 1974 to 1977, such provinces of Sichuan, Shaanxi and Gansu started the first investigation of the population of the giant panda, which was figured out to be more than 2,400.

- In 1975, as the arrow bamboo in the Minshan Mountain blossomed, over 138 giant pandas died; in 1983, the young arrow bamboo in the Minshan Mountain and the Qionglai Mountain extensively blossomed, which led to a successive death of a total of 141 giant pandas.

- On September 8, 1978, the giant panda named 'Juan Juan' in the Beijing Zoo gave birth to two cubs at a litter, which is the first record of success in artificial insemination for the giant panda in the world.

- In 1978, the '51 Tent' field panda observation station, the first of its kind in the world, was set up in the Wolong Nature Reserve, where Chinese and foreign scientists can conduct studies in the individual habits, population and the staple food — bamboo of the giant panda by means of radio tracking and others, and have made remarkable achievements.

- In 1978, the Natural Science Foundation of China co-operated with the United Nations Endangered Animals Protection Center to set up a joint group for the protection and scientific expedition of the giant panda.

- In 1980, the World Wildlife Fund (WWF) was invited by the Chinese government to participate in the protection of the wild giant pandas in China.

- In June 1980, China Conservation and Research Center for the Giant Panda was set up in the Wolong Nature Reserve.

- In 1980, the Wolong Nature Reserve was included in the "Man and Biosphere Program" sponsored by the United Nations Educational, Scientific and Cultural Organization.

- 1980年，成都动物园在世界上首次采用冷冻精液繁殖成活大熊猫。

- 1983年，世界自然基金会（WWF）确定当年为"中国熊猫年"。

- 1984年，洛杉矶奥运会前夕，大熊猫"永永"和"迎新"被租借到美国巡展3个月，它们成了第一对以巡展方式走出国门的大熊猫。

- 1985年—1988年，在四川、陕西和甘肃三省开展了全国第二次大熊猫调查，经统计全国野外大熊猫共有1100只。

- 1987年3月，成都大熊猫繁育研究基地建立。

- 1990年，北京第十一届亚洲运动会组委会确定以大熊猫为本届亚运会吉祥物。

- 1990年，成都动物园首次育活了该园大熊猫"庆庆"所产的双胞胎（在这以前只能育活1崽）。

- 1994年，成都大熊猫繁育研究基地的两只大熊猫首次以"科研交流大使"的身份，旅居日本白浜町动物园，拉开了大熊猫出国合作研究的序幕。

- 1999年3月11日，中央政府赠送给香港特别行政区一对大熊猫"佳佳"和"安安"。

- 1999年—2003年，全国第三次大熊猫调查，结果显示我国野外大熊猫有1596只。

- 2005年9月开始制作，2008年5月上映的电影《功夫熊猫》将大熊猫搬上了好莱坞的银幕，引起强烈的反响。

- 2005年11月11日，北京奥组委确定大熊猫为2008年北京奥运会吉祥物之一。

- 2006年4月28日，四川卧龙自然保护区首次将一只人工圈养的大熊猫"祥祥"放归到野外独立生活，正式启动了"圈养大熊猫野外放归工作"，标志着人类通过科技努力，壮大大熊猫野生种群、促进种群基因交流、逆转大熊猫濒危宿命取得了里程碑意义的突破。

- 2006年7月12日，联合国教科文组织第三十届世界遗产大会将四川以卧龙自然保护区为核心的大熊猫栖息地列入《世界遗产名录》。

- 2007年8月13日，全球首对全人工授精大熊猫龙凤胎在成都大熊猫繁育研究基地诞生。

2008年5月12日，一场百年不遇的大地震突发在岷江重镇——汶川，大熊猫的安危牵动着亿万人的心。四川卧龙自然保护基地61只大熊猫被成功转运到雅安碧峰峡、成都大熊猫繁育研究基地和北京、武汉、广州、云南等地的动物园寄养。
On May 12, 2008, the strongest earthquake in a century happened in Wenchuan County, Sichuan Province, to which the Wolong Nature Reserve is close, so hundreds of millions of people showed much concern for the safety of the giant pandas there. 61 giant pandas in the reserve were then sent to Bifengxia in Ya'an, the Chengdu Research Base of Giant Panda Breeding, and the zoos in Beijing, Wuhan, Guangzhou and Yunnan Province respectively.

当明星大熊猫遇到明星成龙，一场邂逅化为永恒的爱心。2009年"5·12"汶川地震一周年之际，在成都大熊猫繁育研究基地，国际巨星成龙先生捐资100万元人民币认养了大熊猫"成成""龙龙"，并出任"成都熊猫大使"。
The chance meeting of giant panda stars with the film star Jackie Chan means an everlasting love. On May 12, 2009, the first anniversary of the Wenchuan Earthquake, Jackie Chan, a world-renowned film star from Hong Kong, donated RMB1 million to the Chengdu Research Base of Giant Panda Breeding for raising the giant pandas 'Cheng Cheng' and 'Long Long', and took up the post of 'Giant Panda Ambassador of Chengdu'.

美国小女孩米切尔把自己攒存的2400美元零用钱，捐给了亚特兰大动物园里的大熊猫。

Michelle, an American girl, donated her pocket money of $2,400 to the zoo in Atlanta for the giant pandas there.

它是举世无双的宝贝，全世界的人只要想到它，就会想到中国；它是天生的外交明星、运动明星，是人类的吉祥象征。它与运动结下不解之缘：它是1990年北京亚运会的吉祥物；在1992年巴塞罗那奥运会开幕当天出生的大熊猫宝宝，由前国际奥委会主席萨马兰奇亲自以当届奥运会的吉祥物名字"科比"为它命名；最为引人瞩目的是，它成为2008年北京奥运会的吉祥物之一"晶晶"。（图：成都"国际熊猫节"庆典）

The giant panda is a peerless animal in the world, the mention of which will remind people of China. It is inborn diplomat and sports star, as well as an auspicious totem of mankind. It is tightly bound to sports: it was the mascot of the 1990 Beijing Asian Games; Juan Antonio Samaranchi, former President of the International Olympic Committee, named the giant panda born on the day of the opening ceremony of the Barcelona Olympics after 'Kobe', the name of the mascot of the Olympics in 1992; one of the most striking examples was the giant panda 'Jing Jing' that became one of the mascots of the 2008 Beijing Olympics. (The picture shows the celebration of the 'International Panda Festival' held in Chengdu)

- In 1980, the Chengdu Zoo became the first of its kind in the world to succeed in breeding the giant panda by means of artificial insemination with frozen semen.
- In 1983, the World Wildlife Fund (WWF) designated the year as the 'Year of the Giant Panda of China'.
- In 1984, just before the Los Angeles Olympics was held, the giant pandas 'Yong Yong' and 'Ying Xin' were loaned to be on display in the United States for three months. They were the first giant panda pair going abroad by way of exhibition tour.
- From 1985 to 1988, such provinces of Sichuan, Shaanxi and Gansu started the second investigation of the population of the giant panda, the wild of which reached 1,100 all over the country.
- In March 1987, the Chengdu Research Base of Giant Panda Breeding was set up.
- In 1990, the organizing committee of the 11th Asian Games held in Beijing adopted the giant panda as the mascot of this Asian Games.
- In 1990, the Chengdu Zoo succeeded in breeding the twins produced by the giant panda 'Qing Qing' for the first time (previously only one cub could survive).
- In 1994, two giant pandas from the Chengdu Research Base of Giant Panda Breeding arrived and began to live in the Shirahama Zoo in Wakayama of Japan as 'ambassadors of scientific exchanges', which was a prelude to the joint research by means of Chinese giant pandas going abroad.
- On March 11, 1999, the Central People's Government of China presented Hong Kong a pair of giant pandas 'Jia Jia' and 'An An' as a gift.
- From 1999 to 2003, the third national investigation of the population of the giant panda was conducted, the result of which showed that the wild giant pandas in China came to a total of 1,596.
- Starting to be made in September 2005 and shown in May 2008, the film *Kung Fu Panda* put the giant panda on the Hollywood screen, which evoked a strong response worldwide.
- On November 11, 2005, the organizing committee of the 2008 Olympics designated the giant panda as one of the mascots of the Beijing Olympics.
- On April 28, 2006, the Wolong Nature Reserve sent a giant panda 'Xiang Xiang' bred in captivity to the wild to live for the first time, formally starting the 'Program of Sending Captive Giant Panda to the Wild'. The program is aimed at increasing the wild panda population, boosting the gene exchange among the giant panda population, and saving the giant panda from extinction with the help of science and technology.
- On July 12, 2006, the giant panda habitat with the Wolong Nature Reserve in Sichuan Province at its core was included in the World Heritage List in the 30th World Heritage Convention held by the United Nations Educational, Scientific and Cultural Organization.
- On August 13, 2007, the world's first pigeon pair of giant pandas completely by artificial insemination was produced in the Chengdu Research Base of Giant Panda Breeding.

- 2007年9月12日，中国国家林业局新闻发言人公开宣布，中国将不再向外国政府赠送大熊猫。

- 2008年3月，由中国科学家发起，加拿大、英国、美国、丹麦等国科学家联合参与的国际"大熊猫基因组研究"项目启动，绘制大熊猫基因组序列图谱是该项目的第一部分。

- 2008年10月11日，深圳华大基因研究院宣布世界首张大熊猫基因组序列图谱绘制完成。

- 2008年12月23日，四川卧龙自然保护区的大熊猫"团团"和"圆圆"前往台湾台北木栅动物园定居。

- 2011年—2015年，第四次全国大熊猫调查完成。调查显示，全国野生大熊猫总数为1864只。

- 2012年，全球首只母兽带崽培训的人工繁育大熊猫"淘淘"在四川省石棉县栗子坪自然保护区放归自然。

- 2013年11月6日，大熊猫"张想"被放归到四川省石棉县栗子坪自然保护区。这是全球首只放归野外的圈养雌性大熊猫。

- 2014年，大熊猫种群重建正式启动，旨在完成华蓥山大熊猫重引入可行性专家论证及放归适应基地建设。

- 2017年3月，圈养大熊猫"草草"与野生大熊猫实现自然交配，标志着全球范围内大熊猫首次野外引种试验取得初步成效。

- 2018年10月29日，大熊猫国家公园管理局在四川成都成立。

- 2019年1月15日，大熊猫国家公园四川省管理局在成都市、绵阳市、雅安市、广元市、阿坝藏族羌族自治州、德阳市、眉山市的7个管理分局正式成立。

- 2019年1月，中国第一部以大熊猫为主要记述内容的《四川省志·大熊猫志》正式公开发行。6月，四川省地方志工作办公室、四川省林业和草原局联合编著出版《大熊猫图志》。

- ……

（周孟棋、廖继全 辑）

成都儿童展示大熊猫绘画作品。
Children in Chengdu are displaying their drawings of the giant panda.

2008年12月23日，在台湾同胞的期待中，大熊猫"团团""圆圆"终于成功赴台，把爱传递到海峡彼岸。（图：台北市立动物园园长叶杰生亲赴四川迎接大熊猫入台）
On December 23, 2008, the two giant pandas 'Tuan Tuan' and 'Yuan Yuan' arrived in Taiwan, an event the compatriots in Taiwan had been longing for. (The picture shows that Ye Jiesheng, Head of the Taipei Zoo, went to Sichuan in person and received the two giant pandas into Taiwan)

2016年9月4日，世界自然保护联盟（IUCN）将中国大熊猫受威胁等级从"濒危"降为"易危"。
On September 4, 2016, the International Union for Conservation of Nature (IUCN) adjusted the danger level of the giant panda from the previous 'endangered' to present 'vulnerable'.

2018年10月29日，大熊猫国家公园管理局在成都揭牌。（摄影：谢华萍）

On October 29, 2018, the Giant Panda National Park Administration was established in Chengdu, Sichuan Province. (Photo by Xie Huaping)

2018年12月27日，龙溪–虹口放归。

On December 27, 2018, the giant pandas were released to the wild in Longxi-Hongkou.

2019年《四川省志·大熊猫志》《大熊猫图志》中英文版出版发行

In 2019, *The Chronicle of the Giant Panda* and *The Illustrated Chronicle of the Giant Panda* were published in Chinese and English.

- On September 12, 2007, the State Forestry Administration of China announced that the Central People's Government would not present any giant panda to any foreign government as a gift any more.

- In March 2008, initiated by Chinese scientists, partaken by the scientists from Canada, the United Kingdom, America and Denmark, the 'International Giant Panda Genome Project' kicked off. Sequencing the genome of the giant panda would be part of the project.

- On October 11, 2008, BGI-Shenzhen announced that the first genome sequencing of the giant panda in the world was completed.

- On December 23, 2008, the giant pandas 'Tuan Tuan' and 'Yuan Yuan' from the Wolong Nature Reserve traveled to Taiwan and settled down in the Taipei Zoo.

- From 2011 to 2015, the fourth national investigation of the population of the giant panda was conducted, the result of which showed that the wild giant pandas across China came to a grand total of 1,864.

- In 2012, the artificially-bred giant panda 'Tao Tao' returned to the nature in the Liziping Nature Reserve of Sichuan Province. This is the first giant panda of its kind in the world who was given birth to in a specific field environment and taught the surviving skills in the wild by a female giant panda, without any human intervention.

- On November 6, 2013, the giant panda 'Zhang Xiang' was released to the Liziping Nature Reserve, Shimian County, Sichuan Province of China. This is the world's first captive male panda released into the wild.

- In 2014, the reconstruction of the giant panda population was officially launched, which aims to complete the expert argumentation on the feasibility of re-introducing the giant panda to the Huaying Mountain and the construction of an adaptation base for the return of the giant panda to the nature.

- In March 2017, the captive giant panda 'Cao Cao' completed natural mating with a wild giant panda, which marked a preliminary success in the world's first introduction of the giant panda from the wild.

- On October 29, 2018, the Giant Panda National Park Administration was established in Chengdu, Sichuan Province.

- On January 15, 2019, the Giant Panda National Park Administration in Chengdu of Sichuan Province officially established its seven branches in Chengdu, Mianyang, Ya'an, Guangyuan, A'ba Tibetan & Qiang Autonomous Prefecture, Deyang and Meishan respectively.

- In January 2019, *The Chronicle of the Giant Panda* in *The Chronicle of Sichuan Province*, the first chronicle featuring giant pandas in China, was officially released. In June, *The Illustrated Chronicle of the Giant Panda* jointly edited by the Sichuan Provincial Local Chronicles Office and the Sichuan Provincial Forestry and Grassland Bureau was published.

- …………

(Compiled by Zhou Mengqi and Liao Jiquan)

大熊猫：我钟爱的黑白精灵
The Giant Panda: My Beloved Black and White Elf

　　一个物种的价值往往如钻石般具有多面性。大熊猫的价值绝不仅体现于生物学的地位，还包含了美学、生态学，甚至哲学等多样的内涵。这些内涵价值在全世界范围内共通，因而世界自然基金会（WWF）选择大熊猫的形象作为其标志，这也正是这一物种多元文化价值的最好体现。

　　大熊猫是中国的国宝，更是全人类珍视的自然历史的宝贵遗产，具有不可替代性。在很多地方，大熊猫实际上已经被认为是中国的符号元素，代表了中国人民热爱和平、友善待人的品质，体现了中国人坚强的意志以及对环境变化的特殊适应力。不仅如此，这种化石级物种，充分体现了人类与自然和谐共处的永恒主题。

　　因此，我们深爱着大熊猫。而作为"熊猫故乡"的成都人，我对大熊猫的爱，都凝聚在我的照片中。自1992年以来，我持续关注和拍摄大熊猫及其栖息地，自然比常人更接近这个物种。我喜欢用镜头去探索这个物种的神秘世界。大熊猫遵循"物竞天择"之道，于数以万类物种中求存活，是适者生存的典范。历史上的大熊猫生活在更广的地域，有更大的种群与数量。现代，由于地球气候、生态环境日趋恶化，野生大熊猫的生存形势愈加严峻，以至于后来成为地球上的濒危物种，目前野生大熊猫也只剩下不足2000只。值得庆幸的是，近几十年来，国家投入了大量人力、物力来保护大熊猫，在大熊猫繁育与种群保护中取得了可喜的成就。

　　今天，大熊猫的可爱与稀有，使其成为全世界的宠儿，其无与伦比的亲和形象，打破了文化、政治、信仰的疆界，成为整个人类所共有的文化符号，成为抚慰人们心灵的天使、友谊与和平的化身。

　　30多年坚持不间断地拍摄和积累，是个艰辛又漫长的过程，我因此放弃了很多休息时间和与亲朋团聚的机会。多年来，我在拍摄大熊猫的生活及栖息地的过程中尽管屡有惊险，历经种种艰难困苦，但得到更多的是喜悦与信心。每当我看到这些长年积累的光影资料，拍摄中的每段过程、每个场景都历历在目。如今，我依然坚守信念，我会用我的镜头，用自己喜爱的方式一如既往地关爱大熊猫，关注与大熊猫和人类生存息息相关的大自然，关注我们的地球家园以及生活其上的人们。

　　从大熊猫的生存环境到生存现状，从它的生活习性到求偶繁殖，从人工保护到放归自然行动等等，几十年来，我辗转于四川卧龙、雅安、王朗，陕西佛坪、周至等大熊猫自然保护区，拍摄了数以万帧的大熊猫及其栖息地的照片。我希望用这些珍贵的镜头，一方面为生物学家提供具有科研价值的山野调查素材，另一方面为广大读者奉献更多了解、认识、保护大熊猫及其栖息地的科普教材。通过这些照片，我们更直观形象地看到大熊猫除了在动物园被人工饲养之外，另一番更接近生命本真意义的常态。

周孟棋

2021年6月

The value of a species is often judged as being profound and multi-faceted like that of a diamond. The value of the giant panda is not only reflected in its biological status, but also in its connotations in aesthetics, ecology and even philosophy. Such connotations and value have been shared worldwide, so the World Wide Fund for Nature (WWF) has chosen the image of the giant panda into its own logo, which is the just best demonstration of the multicultural value of this species.

The giant panda is treated as a national treasure of China and also an irreplaceable heritage of the natural history cherished by all human beings. In many foreign regions, the giant panda has been recognized as a symbol of China, representing the peace-loving and friendly qualities of the Chinese people, as well as their strong will and exceptional adaptability to environmental changes. Besides, this fossil of species fully embodies the humankind's eternal pursuit of harmonious coexistence with nature.

Therefore, we deeply love the giant panda. And as a native of Chengdu, the "hometown to the giant panda," I try to express my love of it in my photos. Since 1992, I have been following to photograph the giant pandas and their habitats, so I am much closer to this species than ordinary persons. I love exploring the mysterious world of this species through my lens. The giant panda is a typical survivor of the fittest, best illustrating the principle of natural selection and thus surviving among tens of thousands of species. Historically, they used to live in a much wider area with a larger population. However, due to the deteriorating climate and ecological changes in modern times, the survival for wild giant pandas has become more critical, which has actually caused them to be an endangered species on earth. Currently, there are less than 2,000 giant pandas left in the wild. But thankfully, in recent decades, China invested a lot of human and material resources to protect the giant panda, and has made gratifying achievements in breeding and conservation of this species.

Today, due to its adorableness and rarity, the giant panda has been the favorite of the world. By its incredibly affectionate image, it has broken the boundaries of culture, politics and beliefs, and turned to be a cultural symbol shared by all mankind, an angel that soothes our hearts, and an embodiment of friendship and peace.

It is an arduous and lengthy process to photograph for more than 30 years persistently. As a result, I have given up a lot of spare time and opportunities to reunite with family and friends. Over the years, the process of photographing the life and habitats of the giant pandas has often met with too many thrills and hardships, but more with joy and confidence. The moment I look at the photos taken over the years, the details of each process vividly come up to me. Until now, I have held fast to my faith. I will use my lens as well as my means of caring for the giant panda as I always did, to focus on the nature that is closely related to the survival of both the giant panda and human beings, and on the earth and the people living on it.

In the past few decades, I have moved around such giant panda nature reserves as Wolong, Ya'an, Wanglang in Sichuan Province, and Foping, Zhouzhi in Shaanxi Province, taking tens of thousands of photos of the giant pandas and their habitats concerning their living environment, current survival status, living habits, courtship and breeding, captive protection and releasing to the wild. I hope that these precious shots would not only provide biologists with scientific research materials, but also readers with popular science for understanding and protecting the giant panda and its habitat. Through these photos, we are expected to see more clearly how the giant panda actually lives in the wild, as well as kept in captivity in zoo.

Zhou Mengqi

June , 2021

把永恒在瞬间收藏

The eternity is represented in moments

　　周孟棋，中国著名大熊猫摄影家。中国摄影家协会会员、中国新闻摄影学会会员、中国艺术摄影学会会员。先后获中国摄影家协会"德艺双馨优秀会员""抗震救灾优秀摄影家"；2020年获（第12届）中国摄影年度十佳人物等荣誉。

　　持续30余年关注大熊猫，在拍摄大熊猫及其栖息地的过程中，捕捉到很多大熊猫的珍贵瞬间。国内外出版社为其出版摄影专集10余部，包括《我是你的朋友大熊猫》《中国大熊猫》《可爱的大熊猫》《我是大熊猫》《可爱熊猫》《我爱熊猫》等。其中《中国大熊猫》被评为全国美术图书金奖，入选"经典中国"国际出版工程，分别在中、英、意、德、日等国家以不同语种出版。《我是大熊猫》被评为世界华文科普佳作奖，2015年、2016年被国务院新闻办选定为中国驻外使馆交流读物。大熊猫作品《哥俩》荣获迎G20全球摄影大赛金奖。其作品先后赴瑞士、日本、美国、西班牙、法国、英国、德国、比利时、土耳其、澳大利亚、泰国、俄罗斯等国举办个展和巡展，受到普遍欢迎。迄今已在国内外各类报纸杂志上发表摄影作品、文章5000多幅（篇），400多幅摄影作品入选省、全国和国际影展。

　　2018年8月，在北京"首届中国大熊猫国际文化周"上被四川省人民政府评为首批"大熊猫文化全球推广大使"；2020年12月被评为四川省大熊猫保护突出贡献奖先进个人。

Zhou Mengqi, a well-known photographer of the giant panda in China, is a member of China Photographers Association, China Photojournalists Society, and China Artistic Photography Society. He has been successively granted such honors as the 'Outstanding Member of Professional Excellence and Moral Integrity', the "Excellent Photographer for Earthquake Relief" by China Photographers Association, and etc. He won the award for the 12th Annual Top Ten Chinese Photographers in 2020. With the attention paid to the giant panda for nearly 30 years, Zhou has captured numerous precious moments of giant pandas when photographing them and their habitats. He has thus far published more than 10 photo albums in domestic and foreign publishing houses such as *I Am Your Friend Giant Panda*, *The Giant Panda of China*, *Adorable Giant Panda*, *I Am the Giant Panda*, *Lovely Panda*, and *I Love Panda*. Among them, *The Giant Panda of China* won awarded the 'National Gold Award for Art Books' and selected into the 'China Classics International' publishing project, published in China, the UK, Italy, Germany, Japan and others in respective language. The album *I Am the Giant Panda* obtained the 'Award of World Chinese Science Popularization Excellent Work', and was selected as the reading material for communication in the embassies in foreign countries by the State Council Information Office of the People's Republic of China in 2015 and 2016 respectively. Besides, the work *Brothers* about the giant panda was honorable enough to win the 'Gold Award of G20 International Photography Contest'. Zhou's works have been displayed in solo exhibitions and itinerant exhibitions in such countries as Switzerland, Japan, the United States, Spain, France, the UK, Germany, Belgium, Turkey, Australia, Thailand and Russia, winning widespread popularity. As of now, Zhou has published over 5,000 photography works and articles in newspapers and magazines of all kinds at home and abroad, and more than 400 of his works have been selected into the provincial, national and international photo exhibitions.

In August 2018, Mr. Zhou Mengqi was awarded among the first 'Global Promotion Ambassadors of Giant Panda Culture' by Sichuan Provincial People's Government at 'First China Giant Panda International Culture Week' held in Beijing, and in December 2020, he was awarded the 'Advanced Individual of Outstanding Contribution Award for the Giant Panda Protection' of Sichuan Province.

图书在版编目（CIP）数据

中国大熊猫：汉、英 / 周孟棋摄影. —— 成都：四川美术出版社, 2020.12
ISBN 978-7-5410-9780-5

Ⅰ.①中… Ⅱ.①周… Ⅲ.①大熊猫 – 图集 Ⅳ.
①Q959.838-64

中国版本图书馆CIP数据核字(2021)第105333号

中国大熊猫
ZHONGGUO DAXIONGMAO
周孟棋 摄影

科学顾问	胡锦矗　张和民　张志和
撰　　文	樊　奔
英文翻译	陈　伟　李爱华
英文校对	白　雅

责任编辑	聂　平
责任校对	陈　玲
责任印制	黎　伟
装帧设计	张苏坡　周孟棋
封面设计	谢慧婕
设计制作	成都华桐美术设计有限公司

出版发行	四川美术出版社
	成都市锦江区金石路239号
印　　刷	成都中嘉包装印刷有限公司
成品尺寸	230mmx275mm
印　　张	16
字　　数	300千
版　　次	2021年6月第1版
印　　次	2021年6月第1次印刷
书　　号	ISBN 978-7-5410-9780-5
定　　价	580.00元

The Giant Panda of China
Photographed by Zhou Mengqi

Advisers: Hu Jinchu, Zhang Hemin and Zhang Zhihe
Author: Fan Ben
English Translators: Chen Wei and Li Aihua
Translation Proofreader: Bai Ya

Responsible Editor: Nie Ping
Responsible Proofreader: Chen Ling
Person in Charge of Printing: Li Wei
Packaging Designers: Zhang Supo and Zhou Mengqi
Cover Designer: Xie Huijie
Designed and Produced by: Hua Tong Art Design Co., Ltd., Chengdu

Publisher: Sichuan Fine Arts Publishing House of Sichuan Publishing Group
　　　　　No.239, Jinshi Rd., Jinjiang District, Chengdu, Sichuan, 610023
Typeset and Printed by: Chengdu Zhongjia Packaging and Printing Co., Ltd.
Size: 230mmx275mm
Printing Paper: 16
Word Count: 300,000 words
Edition: First edition in June, 2021
Printing: First printing in June, 2021
ISBN: 978-7-5410-9780-5
Price: RMB 580.00